21世纪普通高校计算机公共课程规划教材

Android
应用程序设计

李鲁群 张二江 编著

清华大学出版社
北京

内 容 简 介

本书首先介绍 Android 系统的基本概况、Activity、Service、ContentProvider、BroadcastReceiver 以及 Intent 组件通信基本概念，然后用一章篇幅专门讲解 Android 系统的 UI 与线程通信知识的难点，在此基础上介绍了数据存储、传感器数据采集、网络应用开发、地图导航等相关开发知识。内容基础知识与概念表述清晰，教学案例精心设计，实用性强。本书比较适合作高等院校教材，也可供相关专业人士参考。

图书在版编目(CIP)数据

Android 应用程序设计/李鲁群，张二江编著. --北京：清华大学出版社，2015
21 世纪普通高校计算机公共课程规划教材
ISBN 978-7-302-40484-2

Ⅰ. ①A…　Ⅱ. ①李…　②张…　Ⅲ. ①移动终端－应用程序－程序设计－教材　Ⅳ. ①TN929.53

中国版本图书馆 CIP 数据核字(2015)第 129982 号

责任编辑：黄　芝　李　晔
封面设计：何凤霞
责任校对：时翠兰
责任印制：何　芊

出版发行：清华大学出版社
　　网　　址：http://www.tup.com.cn，http://www.wqbook.com
　　地　　址：北京清华大学学研大厦 A 座　　邮　　编：100084
　　社 总 机：010-62770175　　邮　　购：010-62786544
　　投稿与读者服务：010-62776969，c-service@tup.tsinghua.edu.cn
　　质 量 反 馈：010-62772015，zhiliang@tup.tsinghua.edu.cn
　　课 件 下 载：http://www.tup.com.cn，010-62795954
印 装 者：三河市金元印装有限公司
经　　销：全国新华书店
开　　本：185mm×260mm　　印　张：16.75　　字　　数：403 千字
版　　次：2015 年 11 月第 1 版　　印　　次：2015 年 11 月第 1 次印刷
印　　数：1～2000
定　　价：34.50 元

产品编号：062198-01

出版说明

随着我国改革开放的进一步深化，高等教育也得到了快速发展，各地高校紧密结合地方经济建设发展需要，科学运用市场调节机制，加大了使用信息科学等现代科学技术提升、改造传统学科专业的投入力度，通过教育改革合理调整和配置了教育资源，优化了传统学科专业，积极为地方经济建设输送人才，为我国经济社会的快速、健康和可持续发展以及高等教育自身的改革发展做出了巨大贡献。但是，高等教育质量还需要进一步提高以适应经济社会发展的需要，不少高校的专业设置和结构不尽合理，教师队伍整体素质亟待提高，人才培养模式、教学内容和方法需要进一步转变，学生的实践能力和创新精神亟待加强。

教育部一直十分重视高等教育质量工作。2007年1月，教育部下发了《关于实施高等学校本科教学质量与教学改革工程的意见》，计划实施"高等学校本科教学质量与教学改革工程(简称'质量工程')"，通过专业结构调整、课程教材建设、实践教学改革、教学团队建设等多项内容，进一步深化高等学校教学改革，提高人才培养的能力和水平，更好地满足经济社会发展对高素质人才的需要。在贯彻和落实教育部"质量工程"的过程中，各地高校发挥师资力量强、办学经验丰富、教学资源充裕等优势，对其特色专业及特色课程(群)加以规划、整理和总结，更新教学内容、改革课程体系，建设了一大批内容新、体系新、方法新、手段新的特色课程。在此基础上，经教育部相关教学指导委员会专家的指导和建议，清华大学出版社在多个领域精选各高校的特色课程，分别规划出版系列教材，以配合"质量工程"的实施，满足各高校教学质量和教学改革的需要。

本系列教材立足于计算机公共课程领域，以公共基础课为主、专业基础课为辅，横向满足高校多层次教学的需要。在规划过程中体现了如下一些基本原则和特点。

(1) 面向多层次、多学科专业，强调计算机在各专业中的应用。教材内容坚持基本理论适度，反映各层次对基本理论和原理的需求，同时加强实践和应用环节。

(2) 反映教学需要，促进教学发展。教材要适应多样化的教学需要，正确把握教学内容和课程体系的改革方向，在选择教材内容和编写体系时注意体现素质教育、创新能力与实践能力的培养，为学生知识、能力、素质协调发展创造条件。

(3) 实施精品战略，突出重点，保证质量。规划教材把重点放在公共基础课和专业基础课的教材建设上；特别注意选择并安排一部分原来基础比较好的优秀教材或讲义修订再版，逐步形成精品教材；提倡并鼓励编写体现教学质量和教学改革成果的教材。

(4) 主张一纲多本，合理配套。基础课和专业基础课教材配套，同一门课程有针对不同层次、面向不同专业的多本具有各自内容特点的教材。处理好教材统一性与多样化，基本教材与辅助教材、教学参考书，文字教材与软件教材的关系，实现教材系列资源配套。

(5) 依靠专家，择优选用。在制定教材规划时要依靠各课程专家在调查研究本课程教材建设现状的基础上提出规划选题。在落实主编人选时，要引入竞争机制，通过申报、评审确定主题。书稿完成后要认真实行审稿程序，确保出书质量。

繁荣教材出版事业，提高教材质量的关键是教师。建立一支高水平教材编写梯队才能保证教材的编写质量和建设力度，希望有志于教材建设的教师能够加入到我们的编写队伍中来。

21 世纪普通高校计算机公共课程规划教材编委会

联系人：魏江江 weijj@tup.tsinghua.edu.cn

前　言

自从 2007 年 11 月 5 日 Google 发布了 Android 1.0 Beta 操作系统以来，至今 Android 系统已经发展到 Android 5.0(Lollipop)版本。Android 系统已经从原先单一的、仅支持手机的移动操作一跃成长为支持智能手机(Android Mobile Phone)、平板电脑(Android Tablet)、电视(Android TV)、可穿戴设备(Android Wearable)、车载设备(Android Auto)等众多平台的智能操作系统，而且市场占有率非常高。

在 Android 操作系统发展的同时，形成了一条集半导体芯片、手机制造、手机软件、网络运营商、Android 软件市场、开发者和用户的完整价值链体系与产业生态环境，而且在不断成熟壮大。这其中，人才是关键要素之一。

目前国内外人才市场对 Android 开发人才需求巨大，如何让具有一定 Java 开发知识的学生或开发爱好者能迅速掌握 Android 应用开发知识，是我们教育者应该思索和完成的任务。在多年"Java 程序设计"、"Android 程序设计"课程的本科、研究生教学和企业培训工作中我们认识到，虽然"Android 应用开发"所涵盖的内容极其庞大，但其中关键的知识点主要在于 Activity、Service、ContentProvider、BroadcastReceiver、Intent 通信和 UI 与线程通信，只要掌握了这些内容，就可以迅速掌握 Android 程序设计的核心知识。以此为出发点，我们撰写了这本教材。

本教材内容覆盖了 Activity、Service、ContentProvider、BroadcastReceiver、Intent 通信、UI 与线程通信、传感器数据采集、网络应用、地图服务，授课学时可在 32～54 学时(教学+实验)内完成。

由于教材撰写比较仓促，Android 系统的技术更新较快，书中难免存在不当之处，敬请各位同行和开发爱好者指正(请把您的建议发到邮箱：success@shnu.edu.cn)，在今后的工作中，我们会不断完善本教材。

编　者

2015 年 7 月

目录

第1章 Android操作系统概述

本章介绍Android操作系统的基础知识，内容主要涉及Android操作系统的产生、系统特点以及相关应用常识。

本章的学习目标：

- 了解Android操作系统的产生及历史；
- 了解Android相关的Open Handset Alliance组织；
- 了解Android系统的特点与应用；
- 了解Android开发人员的软件获利渠道。

1.1 Android系统简介

Android，中文俗称“安卓”系统，意思是“机器人”（官网 http://www.android.com），该操作系统是一个以Linux为基础的开放源代码的针对移动设备的操作系统。该系统源码由Google成立的开放手持设备联盟（Open Handset Alliance，OHA）管理与开发。

Android操作系统开源、系统内核小、界面友好以及可以针对应用的不同需求对系统优化定制等诸多特性，使得Android系统迅速广泛应用于工业控制、智能家电、智能家居、车联网、物联网等应用领域。

目前比较典型的Android系统应用领域有Android智能手机、Android TV、Android手表、Android可穿戴设备、Android物联网系统、Android车载设备等，其中Android操作系统作为智能手机、平板电脑的移动操作系统，已经成为全球最大的智能手机操作系统之一。随着Android系统以及应用的发展，各种各样的Android设备会渗透到人们日常生活的方方面面，成为消费电子类产品的重要组成部分。

1.2 开放手持设备联盟组织

说起Android操作系统，我们不得不首先介绍一下负责支持Android系统的开发、维护与管理组织——开放手持设备联盟。

图1-1 开放手持设备联盟（OHA）组织

开放手持设备联盟（见图1-1），组织是Google公司于2007年11月5日宣布组建的一个全球性的联盟组织，这一联盟将会支持Google发布的手机操作系统或者应用软件，共同开发和维护Android的开放源代码的移动操作系统和相关应用。该组织有5大类成员：

- 芯片厂商(Semiconductor Companies)；
- 手机制造商(Handset Manufacturers)；
- 移动运营商(Mobile Operators)；
- 软件开发商(Software Companies)；
- 商业组织(Commercialization Companies)。

该组织成员覆盖了从手机芯片设计、手机设计与制造、手机通信网络运营到手机软件开发和销售服务等各方面，OHA提供的合作平台为Android提供了广阔的市场(http://www.openhandsetalliance.com/oha_members.html)，并且已经形成了一条非常完整的商业价值链体系。

1.3 Android操作系统的发展简述

2003年10月，有"Android之父"之称的安迪·鲁宾(Andy Rubin，见图1-2)在美国加利福尼亚州帕洛阿尔托创建了Android科技公司(Android Inc.)，并与利奇·米纳尔(Rich Miner)、尼克·席尔斯(Nick Sears)、克里斯·怀特(Chris White)共同发展这家公司。最初，Android系统主要是作为一个数码相机的操作系统；随后安迪·鲁宾等人发现市场对此需求不大，于是Android系统被改造为一款面向智能手机的操作系统。

2005年8月17日，Google收购了Android科技公司，Android科技公司成为Google的一部分。于是安迪·鲁宾(Andy Rubin)进入Google，并负责Android操作系统的相关工作。

2007年11月5日，在Google的领导下，Google联合84家硬件制造商、软件开发商及电信营运商成立OHA来共同研发维护和改良Android系统，并以Apache免费开放源代码许可证的授权方式，发布了Android的源代码，目的是创建一个更加开放自由的移动电话环境。2007年11月5日，OHA对外展示了其第一个以Linux 2.6为核心基础的Android操作系统的智能手机——HTC T-Mobile G1 (见图1-3)。

图1-2 Android之父安迪·鲁宾

图1-3 HTC T-Mobile G1

为了更好地维护Android操作系统的开发，OHA成立了安卓开源项目AOSP(Android Open Source Project，https://source.android.com/)项目，AOSP包括了智能手机网络和电话协议栈等智能手机所必需的功能，主要提供Android系统移植、系统设计更新等免费服务，并向各大硬件制造商、软件开发商提供灵活可靠的系统升级承诺，并保证向它们提供最新版本的操作系统。

自从 2007 年第一款 Android 操作系统的手机问世以来，伴随着手机硬件制造水平的提升，Google 领导的 OHA 不断对 Android 系统进行改进并推出新的版本。

Android 操作系统的版本命名很有特色，从 Android 1.5 Cupcake 以来，Android 所有版本都用一种小甜点命名，版本命令规律是小甜点首个英文字母按 C、D、E、F、G、H、I、J、K……顺序命名(见图 1-4)。目前最新版本为 Android 5.0 Lollipop。各版本信息见表 1-1，涉及 Android 软件版本的兼容性问题，建议开发者仔细查看相关 Android 操作系统不同版本的详细特性，此处不再赘述。

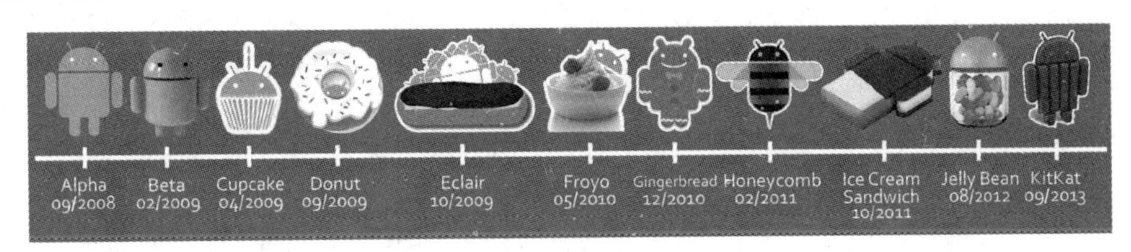

图 1-4　Android 系统发展图示

表 1-1　Android 各版本发布时间

时　间	Android 版本	API 等级
2007 年 11 月 5 日	Android milestone builds-Astro Boy and Bender	无版本号
2008 年 9 月 23 日	Android 1.0	1
2009 年 2 月 9 日	Android 1.1 Petit Four	2
2009 年 4 月 30 日	Android 1.5 Cupcake	3
2009 年 9 月 15 日	Android 1.6 Donut	4
2009 年 10 月 26 日	Android 2.0 éclair	5
2009 年 12 月 3 日	Android 2.0.1 éclair	6
2010 年 1 月 12 日	Android 2.1 éclair	7
2010 年 5 月 20 日	Android 2.2 Froyo	8
2011 年 1 月 18 日	Android 2.2.1 Froyo	
2011 年 1 月 22 日	Android 2.2 Froyo	
2011 年 11 月 21 日	Android 2.2.3 Froyo	
2010 年 12 月 6 日	Android 2.3 Gingerbread	9
2010 年 12 月	Android 2.3.1 Gingerbread	
2011 年 1 月	Android 2.3.2 Gingerbread	
2011 年 2 月 9 日	Android 2.3.3 Gingerbread	10
2011 年 4 月 28 日	Android 2.3.4 Gingerbread	
2011 年 7 月 25 日	Android 2.3.5 Gingerbread	
2011 年 9 月 2 日	Android 2.3.6 Gingerbread	
2011 年 9 月 21 日	Android 2.3.7 Gingerbread	
2011 年 2 月 22 日	Android 3.0 Honeycomb	11
2011 年 5 月 10 日	Android 3.1 Honeycomb	12
2011 年 7 月 15 日	Android 3.2 Honeycomb	13

续表

时　　间	Android 版本	API 等级
2011 年 10 月 19 日	Android 4.0 Ice Cream Sandwich	14
2011 年 10 月 21 日	Android 4.0.1 Ice Cream Sandwich	
2011 年 11 月 28 日	Android 4.0.2 Ice Cream Sandwich	
2011 年 12 月 16 日	Android 4.0.3 Ice Cream Sandwich	15
2012 年 3 月 29 日	Android 4.0.4 Ice Cream Sandwich	
2012 年 7 月 9 日	Android 4.1 Jelly Bean	16
2012 年 11 月 13 日	Android 4.2 Jelly Bean	17
2013 年 7 月 24 日	Android 4.3 Jelly Bean	18
2013 年 10 月 31 日	Android 4.4 KitKat	19
2014 年 10 月 15 日	Android 5.0 Lollipop	20

其中最值得一提的是，Android 5.0 Lollipop（棒棒糖参见图 1-5）可能成为 Android 操作系统发展过程的又一里程碑，它的主要特性如下：

(1) Android 5.0 支持众多硬件平台。系统除了适用于手机、平板电脑外，还适应其他众多平台，如：安卓电视（Android TV）、车载安卓（Android Auto）、可穿戴式安卓（Android Wear）以及健康追踪平台 Google Fit，并提供了相应的 SDK，可以实现不同平台的 Android 5.0 应用无缝链接。

(2) Android 5.0 中 ART（Android Run Time）正式取代 Dalvik 虚拟机。ART 采用了提前编译 Ahead-of-Time compilation（AOT），ART 在应用程序安装时就会编译应用程序，然后他们将只运行已编译过的应用程序，从而改善系统性能。ART 支持 x86、ARM 和 MIPS 架构的 32 位元与 64 位跨平台的运行模式。

(3) Android 5.0 首次提出“质感设计（Material Design）”的概念，使用它设计的界面、图标层次与质感分明，干净利落基于网格的设计布局，灵敏的动画与切换，边距与深度效果，如光线和阴影，看上去好像是画在不同材质上的，材质具有实体的表面和边缘，接缝和阴影意味着提供给你可以真实去触摸的感觉。

图 1-5　Android 5.0 Lollipop（棒棒糖）

(4) Android 5.0 设备可以实现多用户共享。系统可以建立多个用户配置文件，并提供了相应的安全机制来实现与其他人共享手机。

(5) Android 5.0 改善电池节能设计。提供了“项目电压”(Project Volta)的技术来优化改善电池的消耗问题，提供了电池记录(Battery Historian)的新的开发工具追踪耗电的应用程序。

1.4 Android 系统的主要特点

Android 操作系统是比较优秀的移动操作系统，它在如下几个方面表现突出。

1. Android 系统开源、资源丰富

从相关硬件设计操作系统、到应用软件 Android 系统的开源资源非常丰富(见图 1-6)。Android 操作系统使用开放免费代码许可证，Android 的大部分源码以 Apache 开源条款 2.0 发布，剩下的 Linux 内核部分则继承 GPLv2 许可，所有代码为公开免费的。任何厂商都不须经过 Google 和 OHA 授权便可以随意使用 Android 操作系统(但是制造商不能在未授权情况下在产品上使用 Google 的标志和应用程序)，这样，无论底层操作系统到上层的用户界面，还是应用程序都不存在任何阻碍产业创新的专有权障碍。另外，Google 也不断发布问卷征集开发人员和用户意见和评论，来改进 Android 操作系统。随着 Android 系统的不断改进以及应用的日益丰富，Android 系统平台必然会走向成熟，其用户群体数量会不断壮大。

图 1-6 开源的 Android

2. Android 是针对手机优化的移动操作系统

- 系统使用程序框架：支持组件的复用和更换。
- Dalvik 虚拟机：专门为移动设备进行过优化。
- 操作系统直接支持文件与数据库。Android 操作系统支持各类文件的读写，共享操作。另外，它内置高效 SQLite(见图 1-7)小型关联式资料库管理系统来负责存储数据。
- Android 操作系统支持各种网络。Android 操作系统支持所有的网络制式，包括 GSM/EDGE、IDEN、CDMA、EV-DO、UMTS、Bluetooth、Wi-Fi、LTE、NFC 和 WiMAX。支持无线共享，并支持短信和邮件，支持所有的云端信息和服务器信息。其内置的网页浏览器基于 WebKit 核心，并且采用了 Chrome V8 引擎支持 HTML5。
- Android 操作系统支持的媒体种类丰富。操作系统本身支持 WebM、H. 263、H. 264、MPEG-4 SP、AMR、AMR-WB(in 3GP container)、AAC、HE-AAC、MP3、MIDI、Ogg Vorbis、FLAC、WAV、JPEG、PNG、GIF、BMP。同时，Android 操作系统支持 RTP/RTSP(3GPP PSS 和 ISMA)的流媒

图 1-7 Android 内置 SQLite

体、Flash，另外，系统支持语音输入。

- Android 操作系统支持的硬件资源丰富。Android 操作系统支持摄像头、多点电容/电阻触摸屏、GPS、加速计、陀螺仪、气压计、磁强计、键盘、鼠标、USB Disk、专用的游戏控制器、体感控制器、感应和压力传感器、温度计等。Android 操作系统支持多语言。
- 支持 Android 应用程序的组件跨进程相互调用，组件之间无缝集成。Android 系统支持多任务，提供了 Intent 通信以及相应的 Intent Filter 机制，使得开发人员可以将在自己开发的程序与其他应用组件，如本地的联系人、日历、位置信息、邮件通信，完成组件之间、进程之间、跨进程通信，从而实现组件之间的无缝集成。如，开发人员可以将自己开发的程序中的文本、图片，通过 Intent 直接与微信系统组件通信，把文本和图片发给微信好友。

3. Android 有强大的网络后台服务和 Android 软件销售平台

Android 系统有 Google 强大的网络后台服务做支持和 Android 软件销售平台 Google Player 等。Android 系统可以直接与 Google 搜索、Google Play 商店、Gmail、Google 地图、Google 云存储、云计算等后台服务。除此之外，国内微信、金山云盘、百度地图、搜狗地图等第三方软件与服务提供商也提供了 Android 服务 API，使得 Android 移动应用与网络后台服务实现无缝集成。

4. Android 系统可以根据应用定制

Android 系统除了运行在智能手机外，可以根据不同的应用场景定制裁剪系统，目前有 Android 系统支持的平板电脑、Android TV、Android 手表、Android 可穿戴设备、Android 物联网、Android 车载设备等，摩托罗拉、三星、LG、HTC、宏碁、华硕等公司均推出了平板电脑产品，国内的创维、TCL 等厂商已经推出了 Android 智能电视，不久将有更多的智能家电、机顶盒、车载电子设备出现。

1.5 Android 系统结构

Android 系统是基于 Linux 内核的开源系统，从系统的组成的角度来看，Android 平台架构由硬件设备、驱动程序、操作系统内核、程序运行库、运行框架、应用程序等组成，它们的有机结合和协同工作使得 Android 系统得以正常运行。

系统架构如图 1-8 所示，下面由下而上对组成系统各部分的主要组件做以下描述。

1.5.1 Linux 内核层(Linux Kernel)

Android 内核基于 Linux 2.6 内核做了部分修改和增删，是一个增强内核版本，除了修改部分错误(Bug)外，它还提供了用于支持 Android 平台的硬件设备驱动。Android 核心系统实现了安全性、内存管理、进程管理、网络协议栈和驱动模型等功能，Android 操作系统的 Linux 内核层也同时作为硬件和软件栈之间的抽象层。

1. 硬件驱动

Linux 内核层提供了几乎所有手机、平板电脑相关设备的驱动程序，实现系统与各种硬件的通信，如显示屏、摄像头、内存、键盘、无线网络、音频设备、电源等组件。

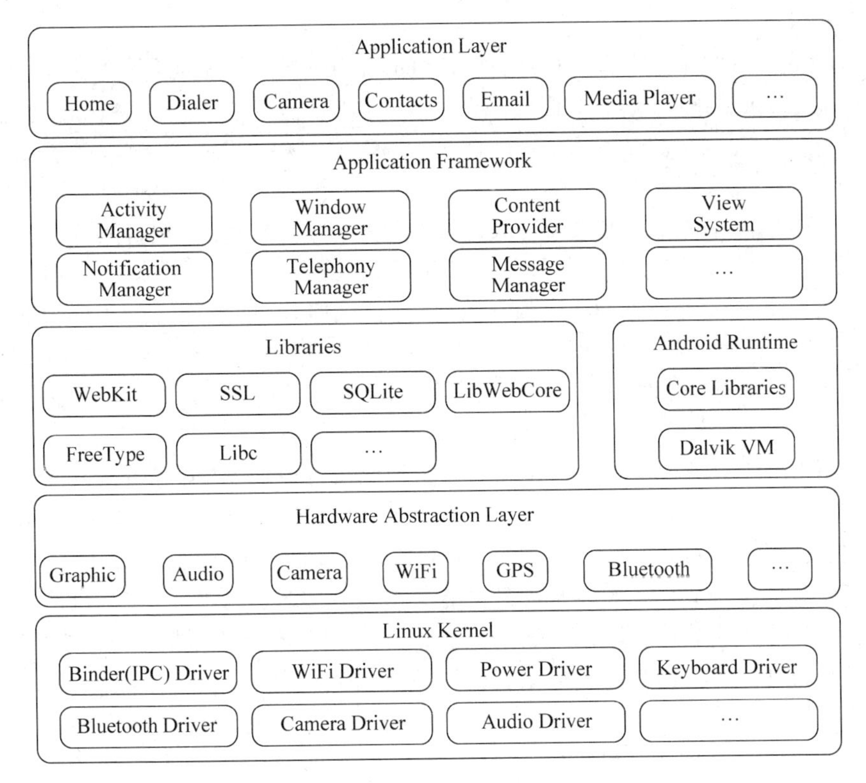

图 1-8　Android 系统平台架构

2. 内存管理

Linux 内核层还提供系统内存管理，实现对所有可用的内存进行统一编码管理，定义一整套内存定位，使用与回收的策略，提供了低内存管理器（Low Memory Killer）策略，Android 系统可以根据系统运行资源情况，自动决定是否需要杀死进程来释放所需要的内存。Linux 内核层还提供了匿名共享内存（ashmem）机制，系统为进程间提供大块共享内存，同时为内核提供回收和管理内存机制。另外，针对 DSP 和某些设备只能工作在连续的物理内存要求，系统内核层提供了 Android PMEM 机制解决了向用户空间提供连续的物理内存区域的问题。

3. 系统进程管理

实现管理进程的创建与销毁，管理进程间的通信，解决与避免死锁问题等。Android 系统的进程间通信基于 Binder 机制实现，一个进程可以非常方便地实现跨进程调用一个进程所提供的功能，并获取返回的执行结果。

4. 文件系统管理

Android 平台采用 Yaffs2 作为 MTD nand flash 文件系统，Yaffs2 使用更小的内存来保存它的运行状态，其垃圾回收机制非常简单快速，在大容量的 NAND Flash 上性能表现尤为突出。

5. 电源管理

Android 电源管理，一个基于标准 Linux 电源管理系统的轻量级的 Android 电源管理驱动，针对嵌入式设备做了很多优化。

6. USB管理

Android的USB驱动是基于Gaeget框架的,USB Gadget驱动是一个基于标准Linux USB gadget驱动框架的设备驱动。

1.5.2 硬件抽象层

Patrick Brady (Google)在2008 Google I/O提出的Android硬件抽象层(Hardware Abstract Layer,HAL)概念和架构图。HAL的目的是把Android框架层与内核隔开。简而言之,就是对Linux内核驱动程序的封装,向上提供接口,屏蔽低层的实现细节。也就是说,把对硬件的支持分成了两层:一层放在用户空间(User Space),一层放在内核空间(Kernel Space);其中,硬件抽象层运行在用户空间,而Linux内核驱动程序运行在内核空间。

之所以使用HAL层,是因为Linux内核源代码版权遵循GNU License,而Android源代码版权遵循Apache License,前者在发布产品时,必须公布源代码,而后者无须发布源代码。如果把支持硬件的所有代码都放在Linux驱动层,那就意味着发布时要公开驱动程序的源代码,而公开源代码就意味着把硬件的相关参数和实现都公开了,在手机市场竞争激烈的今天,这对厂家来说,损害是非常大的。因此,Android才会想到把对硬件的支持分成硬件抽象层和内核驱动层,内核驱动层只提供简单的访问硬件逻辑,例如,读写硬件寄存器的通道,至于从硬件中读到了什么值或者写了什么值到硬件的逻辑中,都放在硬件抽象层中去了,这样就可以把商业秘密隐藏起来了。HAL让Android framework的开发能在不考虑驱动程序的前提下进行发展,也迎合了厂商不希望不公开其硬件驱动源码的要求。

1.5.3 程序库

Android系统程序库(Libraries)包含一些C/C++库,Android系统中不同的组件通过应用程序框架可以使用这些库,常见的核心库如表1-2所示。

表1-2 Android系统中的程序库

库 名 称	功 能 说 明
System C library	一个继承自BSD的标准C系统实现(libc),被调整成面向嵌入式Linux设备
Surface Manage Libraries	提供了管理显示子系统,并且为多个应用程序提供2D和3D图层的无缝融合
OpenGL ES library	3D图形库,用于3D图形渲染,该库可以使用3D硬件加速
FreeType library	提供位图(Bitmap)和矢量(Vector)字体显示库
WebKit library	提供支持Android浏览器和一个可嵌入的Web视图(View)控件
SQLite library	它提供了功能强大的轻型关系型数据库引擎
Media Libraries	基于PacketVideo的OpenCore;该库支持回放和录制许多流行的音频和视频格式,以及静态图像文件,包括MPEG4、H.264、MP3、AAC、AMR、JPG和PNG格式

1.5.4 Android运行库(Android Runtime)

Android运行库包括两大部分:一是核心库,二是Dalvik虚拟机。

核心库提供 Java SE 编程语言核心类库的大多数功能，已经熟练掌握 Java SE 开发的人员，无须再学习这部分知识。

Dalvik 虚拟机是 Google 专为 Android 开发的，Dalvik 虚拟机是基于寄存器，依赖于底层 Posix 兼容的操作系统，它可以简单地完成进程隔离和线程管理。每一个 Android 应用在底层都会对应一个独立的 Dalvik 虚拟机实例，其代码在虚拟机的解释下得以执行。它对内存的高效使用和在低速 CPU 上表现出的高性能，确实令人刮目相看。Dalvik 虚拟机执行的是.dex 结尾的 Dalvik 可执行文件格式，该格式被优化为最小内存使用。Android SDK 提供 DX 编译打包工具，将 Java 编程语言所编译的类转换为.dex 格式。

1.5.5 应用程序框架层

应用程序框架(Application Framework)层定义了一个应用程序运行所必需的全部功能组件，应用程序的架构设计简化了组件的重用；使得开发者自由地享有硬件设备的优势，访问本地信息，运行后台服务，设置警示，向状态栏添加通知等。在 Android 程序框架中，任何一个应用程序都可以发布它的功能块；所有的应用程序在 Android 平台上都是平等的；所有的应用程序与资源都被按类别进行分别管理。底层的所有的应用程序是一组服务和子系统，如表 1-3 所示。

表 1-3 Android 应用程序框架

文件类型	说　明
Windows Manager	窗口管理框架系统动画框架
Packager Manager	包管理服务。资源管理相关类
Notification Manager	以使所有的应用程序在状态栏显示定制的提醒
Activity Manager	它管理应用程序的生命周期，并且提供了一个通用的后台切换栈
Resource Manager	提供对非代码资源的访问，比如本地化的字符串、图形和布局文件
Content Providers	可以使应用程序访问其他应用程序的数据(比如通讯录)，或者共享自己的数据
View UI 可视化	View 可以被用来构建一个应用程序，包括列表、表格、文本框、按钮，甚至可嵌入的 Web 浏览器

1.5.6 应用程序层

应用程序层(Application Layer)的各种软件是用 Java 语言编写的运行在 Dalvik 虚拟机上的程序，如，Android 系统中应用，E-mail 客户端、SMS 短消息程序、日历、地图、浏览器、联系人管理程序等。

1.6 学习 Android 开发先验知识

Android 应用开发主要使用 Java 语言，由于底层是 Linux 系统，支持 C/C++编程。目前有两种应用开发编程途径：第一，基于 ADT 的 JAVA 编程(主流开发方式)；第二，基于 NDK 的 C/C++编程，仅用于硬件或系统相关的开发。

因此，在学习 Android 应用开发之前，学习者必须具备 Java SE 开发核心知识，这包括了熟练掌握 Java 基本数据类型及其特点；Java 分支语句和循环语句的使用；面向对象 Java

程序设计基本知识，类和对象的创建和使用方法，抽象类和接口；继承和实现，对象的多态性；异常的处理；线程的创建与使用、集合类的使用（如 ArrayList、HashMap）；了解网络与数据库相关基本知识；Linux 操作系统基本知识以及 Eclipse 的基本用法。如果学习者不具备这方面的知识，建议学习前先补充学习相关知识。

1.7 Android 开发者如何获利

移动互联网应用已经成为了一个流行趋势，Android 作为一个开放性平台，对手机厂商和软件开发商的吸引力也在持续升高，市场对 Android 开发人员需求巨大，在这种大的背景下，具备 Android 应用程序设计开发知识也成为计算机科学与技术专业学生必备的专业技能之一。学习 Android 程序开发，虽然可以使学习者更“充实”，但是作为 Android 开发者面对日益增长的生活成本开销，不得不面对一个基本的现实问题：“Android 开发者如何获利?”另外，作为一个好的 Android 应用，也必须得到 Android 市场的承认，Android 应用得到市场承认的最简单的评判指标是程序盈利情况。目前 Android 开发者主要获利渠道有三条。

1.7.1 承接项目与产品设计

“做项目”，即：作为公司开发人员或独立开发人员承接 Android 特定领域项目获利；目前国内外有许多网站都会发布相关 Android 项目招标信息。如中国外包网（http://www.010china.com/，见图 1-9）和智诚外包网（http://www.taskcity.com，见图 1-10）。

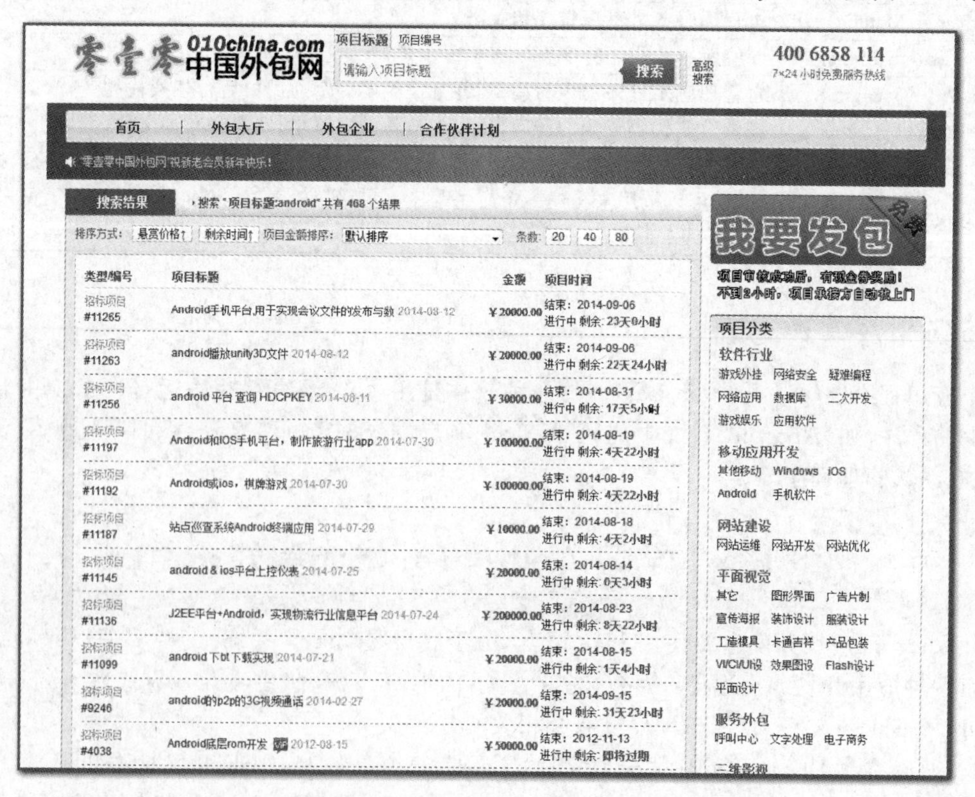

图 1-9　中国外包网（http://www.010china.com/）

图 1-10　智诚外包网(http://www.taskcity.com)

1.7.2　在 Android 软件市场出售 APP

"做产品"，即：作为独立开发人员或自由职业者，将自己的产品上架到 Android 应用市场(如 Google play、安智市场)等，通过在线出售软件方式营利。这种模式通常称为 IAP (In-App Purchase)，俗称"卖软件"，国内用户已经习惯了"免费"使用 Android 软件的模式，对于靠软件收费营利的模式，因为国内购买软件版权用户数量有限，所以很难获利；当然不排除一些 Android 手机游戏软件可以靠玩家买单这种模式营利。

目前较大型的 Android 应用市场有 Google play，这是 Google 官方的应用市场，Android 应用覆盖量自然也是最大的。令人遗憾的是，国内的手机生产厂商由于利益驱动，厂商在自行定制 Rom 过程中，多数厂商已经把原本内置的 Google play 应用删除掉了，所以 Google play 也面临着中国本土特殊环境的尴尬局面；国内比较大型的 Android 应用市场有安卓市场(http://www.hiapk.com)、机锋市场(http://apk.gfan.com)、安智市场(http://market.goapk.com/)、掌上应用汇(http://www.appchina.com/)以及 360 手机助手市场(http://zhushou.360.cn/)。

1.7.3 广告获利

"利用广告联盟获利",开发者将自己开发的 App 提供软件免费下载,开发者与移动广告商签订协议,通过在软件中内嵌广告条,用户在使用软件的过程中,如果点击广告条,开发者便可以由此获利。应用内嵌入广告的模式操作起来相对简单,只要将广告条放置在应用界面的固定位置(通常是顶部或底部)即可。同时,不同的应用还提供了不同的移除广告方式。如,付费后才能下载没有广告的应用版本;也有较为温和的通过求捐助的方式在博得用户同情的基础上提供去广告服务,还有一些资深 Android 开发人员,通过在软件中设计逻辑锁,让用户在安装和使用软件时候,强制用户必须点击广告或安装其他软件后,软件才能解锁移除广告,开发者通过这种方式直接获利。

1.8 Android 手机应用知识拓展

目前 Android 智能手机、平板电脑、机顶盒、手表等应用设备与系统非常普及(读者可在淘宝 http://www.taobao.com/上查询 Android 相关设备产品信息与价格),虽然大多数读者对 Android 系统的使用有了一定的应用体会,但是对 Android 系统专业应用知识还有待充实。以下简单介绍 Android 手机最常见的用户权限问题和刷机问题。

1.8.1 什么是手机 Root

从专业的角度讲,一部 Android 手机,就是类似一部"小电脑",安装的操作系统为类 Linux 操作系统,前台图形界面为我们所看到的用户界面。通常情况下,Android 手机生产厂商为了保障系统的安全、防止用户误操作破坏系统,Android 手机默认用户不具备超级用户 Root 权限。用户可以在用户空间内自行安装、卸载软件,但是没有删除系统内置软件的权限。大多数手机厂商(尤其是许多"山寨"手机生产厂商)利用这一点,在手机售出前内置许多应用,强制用户接受并使用,许多内置恶意手机应用还会直接耗费用户数据流量和话费。

为了解决该问题,许多 Android 手机 Root 工具因此产生。这些 Root 工具通常使用非常简单,软件在手机上运行后,可以直接将当前用户权限提升为超级用户 Root,通常把这个过程成为 Root。一部经过 Root 的手机,用户可以自行卸载删除系统内置的软件或文件,这样可以直接删除系统内置的"垃圾"或恶意软件,但是如果用户误删了手机重要系统文件后,手机可能失去正常手机基本的功能。另外,被 Root 后的手机,如果安装了其他恶意软件,轻则会导致被恶意扣费、耗费用户流量,重则会使手机用户个人隐私泄露、手机银行密码被盗等。所以建议,用户一旦把自己的手机 Root 后,不要从无法信任的网站上下载手机 App 应用。建议 Android 手机用户安装安全可信的正规厂商的手机杀毒软件,来保证手机的安全使用。

1.8.2 什么是"刷机"

与 PC 一样,Android 手机的操作系统也是可以重新安装的。为了简化手机操作系统的安装过程,手机生产厂商通常会把系统做成一个镜像压缩文件,俗称"ROM 文件",用户可以利用该文件恢复系统(类似 PC 克隆恢复系统),通常把这个过程称为“刷机”。

目前,一般手机生产厂商或第三方 ROM 提供服务者(如 http://www.romzhijia.net/)会不定期对所售机型提供新版本的 ROM 及刷机说明,ROM 通常为 update.zip 文件。“刷机”前用户可以把该文件解压缩,查看系统目录\system\app 中有哪些预装软件(通常是预装应用是一个.apk 文件以及对应.odex 文件同时存在)。

用户在“刷机”的时候,一定认准手机型号要与 ROM 类型匹配,若不匹配,“刷机”后可能导致手机无法正常使用,俗称“变砖”(Brick)。由于 Android 系统源码开放,“刷机”同样涉及手机的安全问题,一些黑客和恶意厂商完全可以修改系统预留“后门”、内置恶意收费软件等,然后把手机界面伪装成绚丽界面,吸引一些外行人下载该 ROM 并“刷机”,手机刷机后“中招”,同样会面临个人隐私泄露、恶意扣费、手机银行失盗等安全问题,所以,建议用户不要使用无法信任的 ROM“刷机”。

1.9 本章小结

通过本章学习,已经掌握了 Android 系统的基础知识、Android 系统的商业价值链体系以及 Android 系统的应用常识。

1.10 习题与课外阅读

1.10.1 习题

(1) 开放手机联盟(Open Handset Alliance)组织有哪几类成员?

(2) Android 操作系统支持哪些类型的网络?

(3) Android 开发者获利渠道有哪些?

1.10.2 课外阅读

(1) 访问下列技术网站,了解一下 Android 系统最新技术动态:

- http://www.android.com;
- http://source.android.com;
- http://www.openhandsetalliance.com/;
- https://play.google.com/store。

(2) 访问以下网站了解 Android 硬件市场、软件外包市场以及人才需求市场:

- http://www.taobao.com (搜索 Android 硬件类产品);
- http://www.010china.com/ (Android 软件外包市场);
- http://www.51job.com/ (检索 Android 开发人才的需求)。

第2章 Android开发环境的搭建与使用

本章主要介绍Android开发环境的搭建及使用，并演示一个最基本的Android程序"Hello World"的编写过程，分析Android应用程序的逻辑组成结构，学习Android模拟器的使用，以及应用程序的调试方法。

本章的学习目标：

- 掌握Android开发环境的搭建方法；
- 掌握Android应用程序的逻辑组成结构；
- 学会配置、使用Android模拟器；
- 学会使用ADB、DDMS工具。

2.1 Android开发环境的搭建

Google为Android应用开发提供了Android SDK以及相关集成开发工具(Integrated Development Environment,IDE)，这些开发工具均依赖于Java虚拟机运行环境，所以要使用Android应用开发工具，必须首先安装Java JDK，http://www.oracle.com/technetwork/java/index.html (Java JDK的安装此处不再赘述，建议安装最新版的Java JDK)，然后再去Android SDK的网址下载安装http://developer.android.com/sdk/index.html开发工具，以往开发环境的搭建是安装Android SDK和开发集成环境Eclipse或IntelliJ IDEA，然后再安装集成开发ADT插件来绑定Android SDK与Eclipse或IntelliJ IDEA。为了方便开发者使用，Google直接提供Android SDK Bundle for Windows Eclipse类型版本和Android Studio版本集成开发工具。该版本集成了如下内容：

- Eclipse(或Android Studio)+ADT plugin；
- Android SDK Tools；
- Android Platform-tools；
- The latest Android platform；
- The latest Android system image for the emulator。

下载后，只需解压该工具到一个文件夹中即可使用，非常适合初学者使用。当然开发者也可以不用IDE开发环境直接在Android SDK环境下，手工编写、编译、安装调试Android应用程序，该方法比较耗时费力，不适合初学者，所以不建议使用。

概括起来说，开发环境搭建首先要安装Java JDK；然后安装Android开发工具及Android SDK。安装完后，打开Android SDK Bundle for Windows文件解压缩目录，可以看到如图2-1所示界面。

名称	修改日期	类型	大小
eclipse	2014/4/24/星期四 12:18	文件夹	
sdk	2014/4/1/星期二 11:06	文件夹	
SDK Manager.exe	2014/3/22/星期六 7:50	应用程序	352 KB

图 2-1　Android 开发工具目录

图 2-1 中的 eclipse 和 sdk 目录，分别对应 eclipse 开发工具和 Android SDK，另外还有 SDK Manager. exe 文件，该文件用于管理本地 Android SDK 文件，运行该文件后，该文件直接与网络 Google Android SDK 相关服务建立连接，自动下载文件更新列表，如图 2-2 所示。

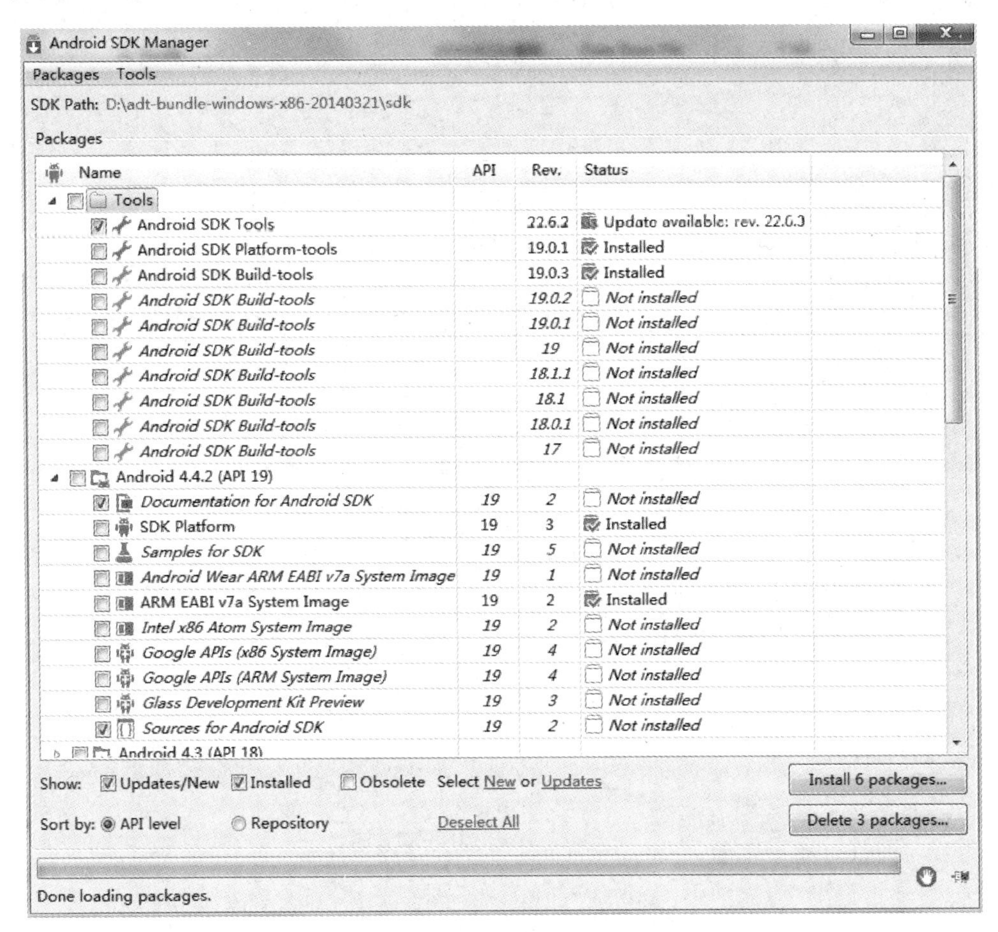

图 2-2　SDK Manager 管理工具

为了方便开发者查找开发相关文档（当然开发者可以在线查找），建议下载开发 API 文档(Documentation for Android SDK)到本地；另外，初学者最好再下载学习 Android 相关经典例子(Sample for SDK)；如果开发者想深入 Android 系统研究，可以下载 Android SDK 源码(Sources for Android SDK)来研究。

进入 Eclipse 目录，运行 Eclipse. exe 文件（本书默认读者已经具备基本 Java 开发知识，所以对 Eclipse 集成开发环境基本使用不再赘述），出现如图 2-3 所示的界面。

单击 Android 虚拟设备管理按钮，可以创建、启动 Android 模拟器（Android

Emulator)。虚拟设备的创建要注意设备的 Android 操作系统的版本号、内存大小、SD 卡容量等够用即可(见图 2-4),如果内存、SD 卡容量设置的过大,将耗费很多 PC 资源,影响程序开发与调试运行速度。

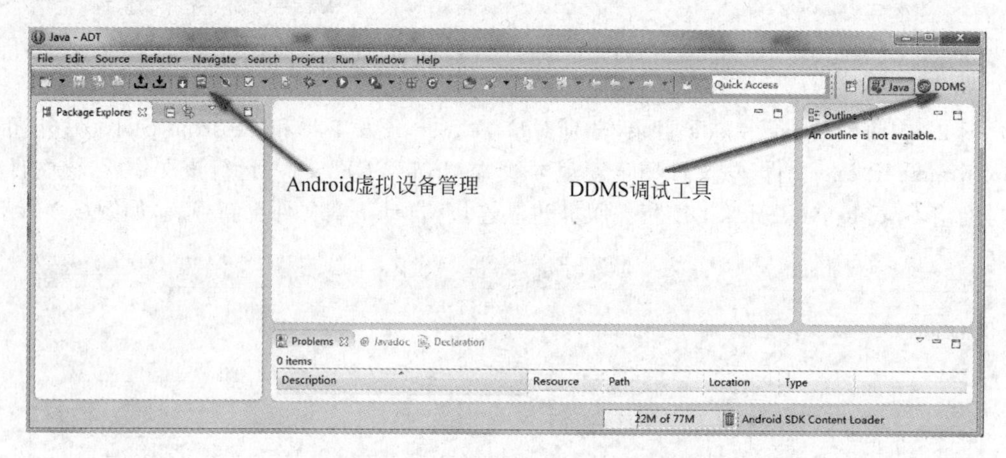

图 2-3 Eclipse Android 开发集成环境

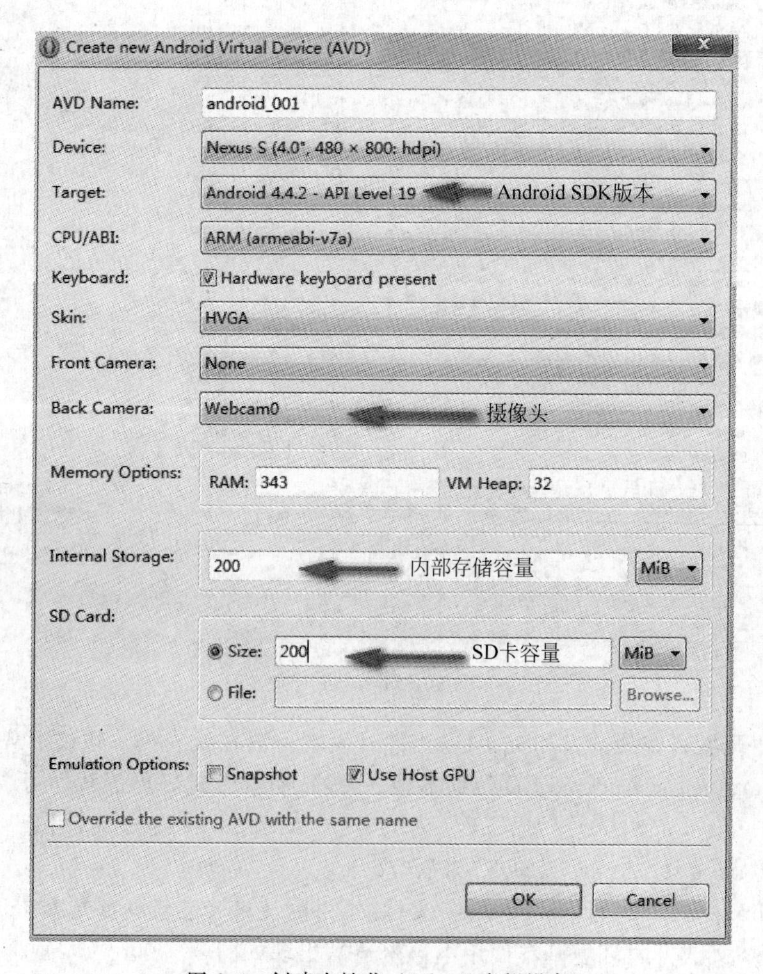

图 2-4 创建个性化 Android 虚拟设备

启动模拟器后，将出现如图 2-5 所示的界面。

图 2-5　Android 模拟器

模拟器相关说明：

(1) 模拟器可将本地 PC 的摄像头映射为模拟器的摄像头，且能创建虚拟 SD 卡。

(2) 如果 PC 内存足够，开发者可以同时启动多个模拟器，每个模拟器/设备实例获取一对序列端口：

- 一个偶数编号的端口用于控制台连接；
- 一个奇数编号的端口用于 adb 连接。

(3) 模拟器可以模拟大部分真实手机的功能，但是模拟器不支持物理传感器（加速度、温度等）和蓝牙功能，涉及这些功能的应用程序通常必须在真机上调试；

(4) 模拟器可以安装虚拟传感器来运行传感器相关程序，如 SensorSimulator (http://code.google.com/p/openintents/wiki/SensorSimulator)，可以模拟的传感器有 accelerometer（加速度计）、compass（罗盘）、orientation（方向传感器）、temperature（温度传感器）、light（光传感器）、proximity（接近传感器）、pressure（压力传感器）、linear acceleration（线性加速度计）、gravity（重力感应传感器）、gyroscope（陀螺仪）和 rotation vector sensors（旋转向量传感器）。

(5) 除了使用虚拟设备外，还可以直接用真实的 Android 物理设备（如 Android 手机、平板电脑等）通过 USB 端口连接 PC 搭建开发环境，Android 模拟器或真实的物理设备与 Android 应用开发与调试计算机通过 ADB(Android Debug Bridge，Android 调试桥）协议进行通信。由于使用真实 Android 物理设备，调试速度远远高于使用虚拟设备，一般在有条件的情况下，建议开发者使用真实设备进行 Android 应用开发(见图 2-6)。

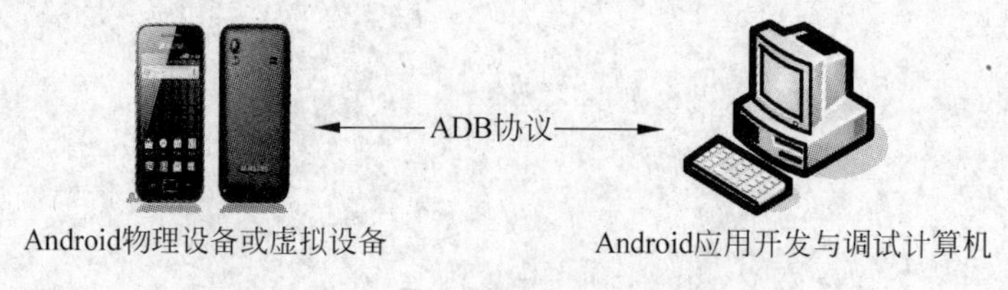

图 2-6　Android 开发环境示意图

至此，已经搭建好 Android 开发环境。下面来编写一个最简单的 Android 应用程序。

2.2　第一个“HelloWorld” Android 程序

Android SDK+Eclipse+ADT 集成开发工具，为开发程序提供了各种向导模板，初学者可以不必编写一行代码，直接利用向导迅速完成一个简单 Android 应用程序设计，如图 2-7～图 2-14 所示。

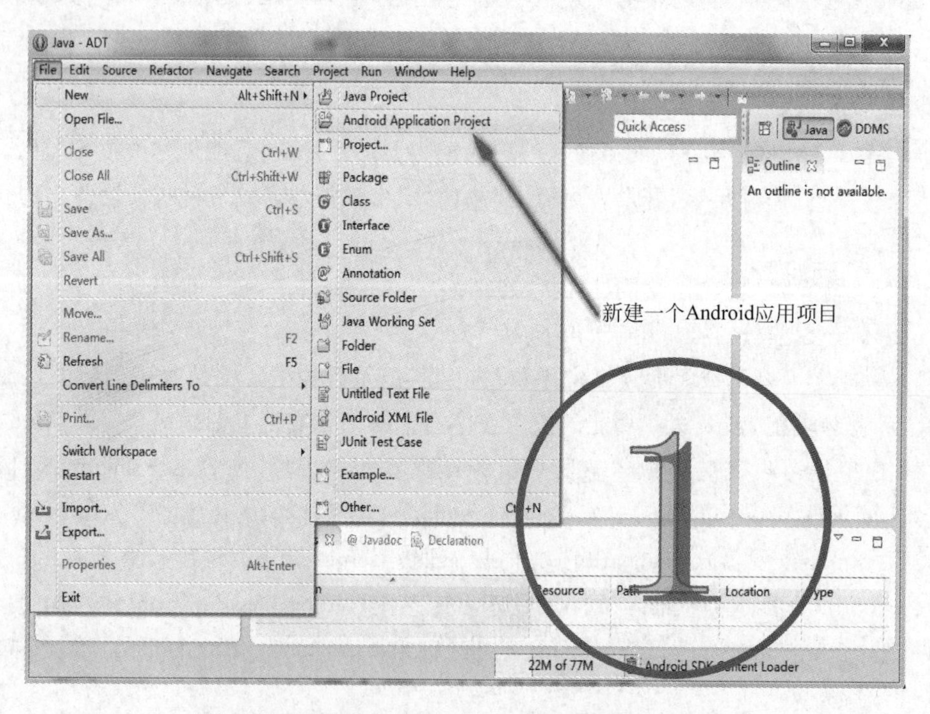

图 2-7　选择 File→Android Application Project 命令

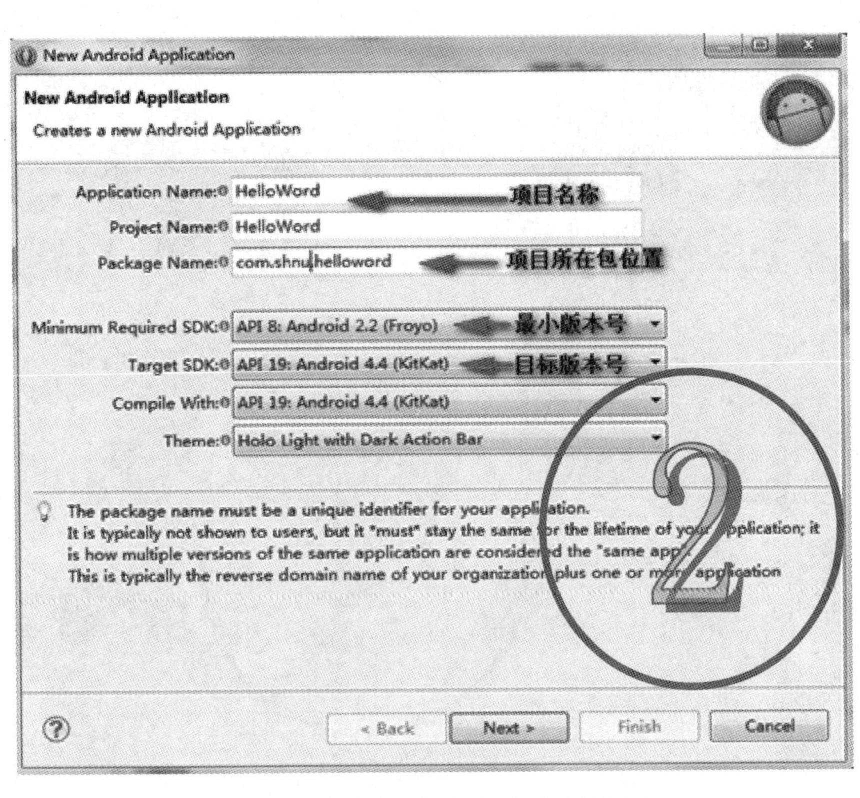

图 2-8　设定项目名称、包名、Android 版本号

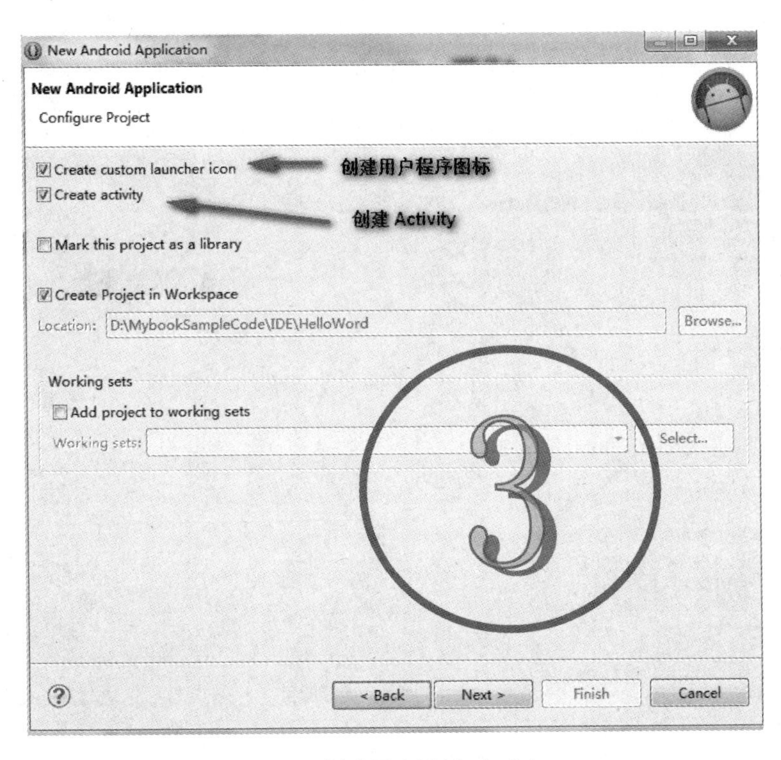

图 2-9　创建用户图标、Activity

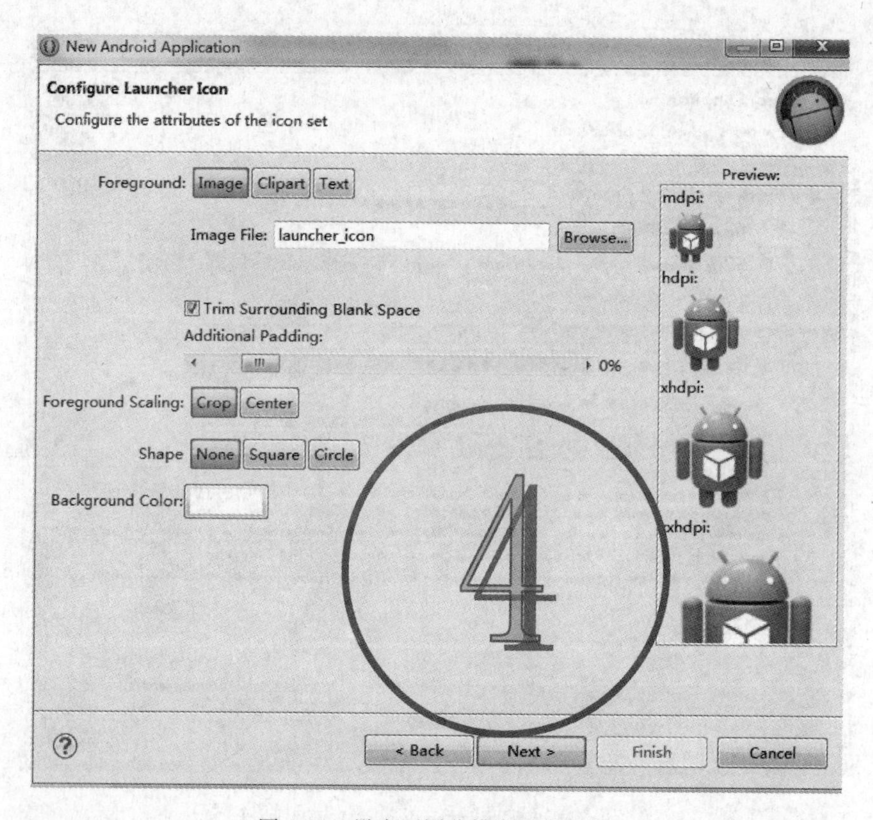

图 2-10 设定不同分辨率程序图标

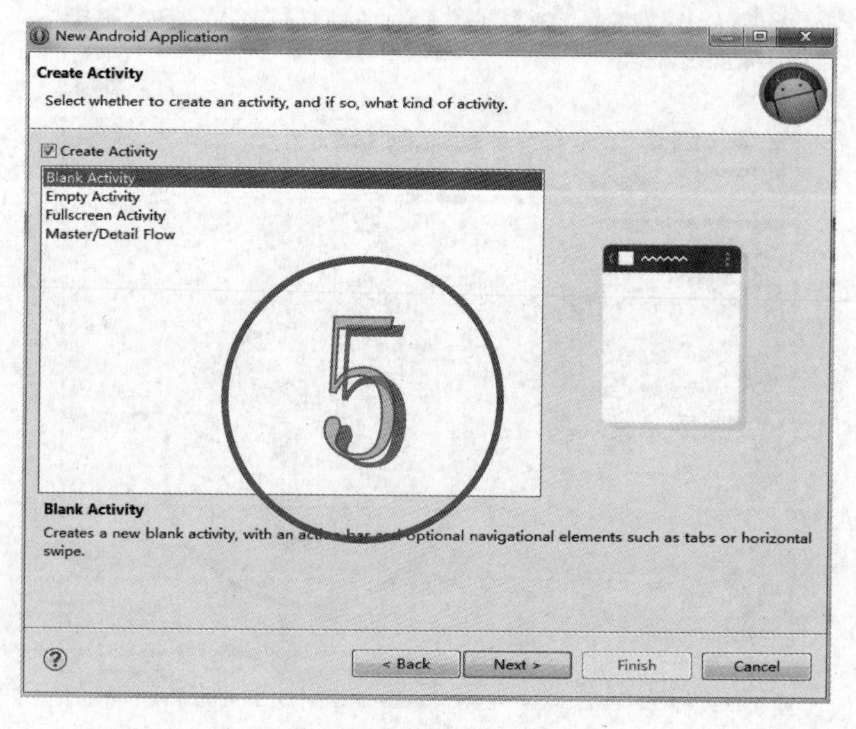

图 2-11 创建空的 Activity

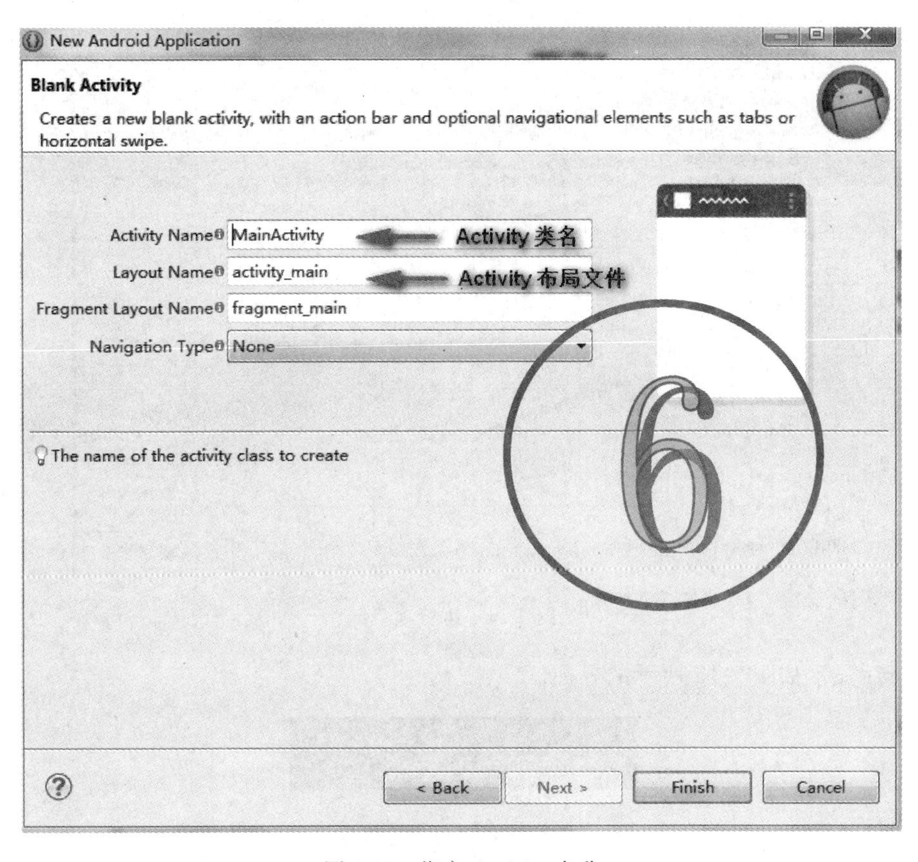

图 2-12　指定 Activity 名称

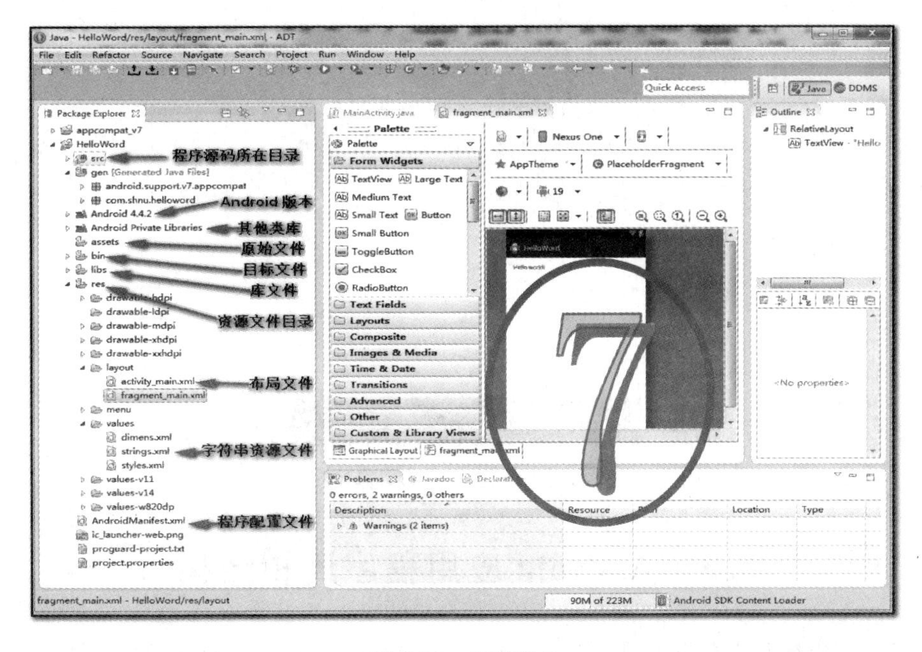

图 2-13　打开项目

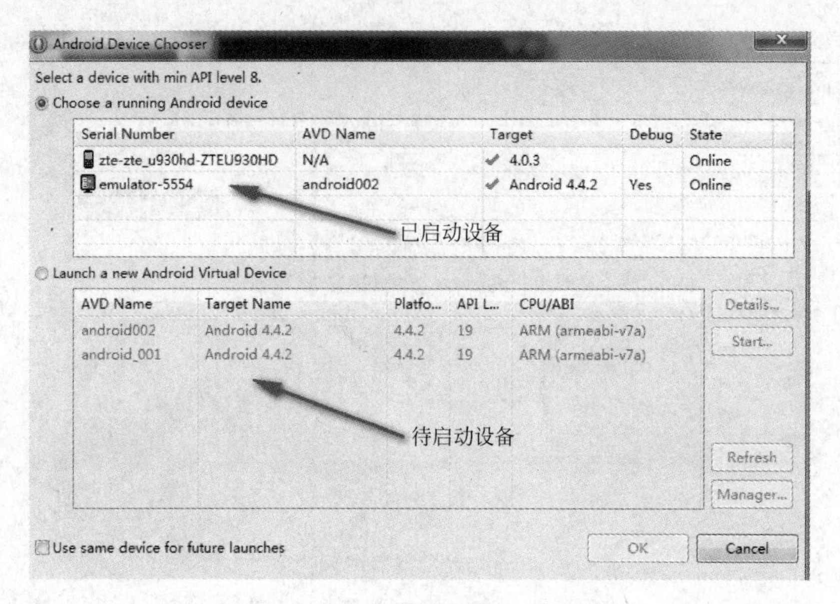

图 2-14　运行项目

运行后显示如图 2-15 所示的界面。

图 2-15　程序运行结果

2.3　Android 应用程序逻辑结构

虽然只一个看似非常简单"Hello World"功能演示程序，整个过程无须手工编写一行代码，事实上是集成开发环境 ADT 插件为我们完成了项目框架、程序资源的创建、程序代码、配置文件的编写。集成开发环境 ADT 插件会生成很多 Android 应用程序必需的目录和文件(见表 2-1 和表 2-2)。下面举例说明。

表 2-1　Android 应用程序文件分类

文件类型	说　明
程序代码	这部分多以 *.java 或 *.aidl 文件存在；主要放置在 src 目录下
资源文件	该类文件包括 *.xml 和 *.txt 等各类文件；这里文件通过 ADT 插件，自动会在 gen 目录下生成对应的 *.java 文件
配置管理文件	该文件定义项目组件、资源、权限等，是系统的核心配置文件，其功能有点类似 Java EE 应用的 web.xml 文件

表 2-2　Android 项目中包含的主要目录内容

目录名称	说　明	
src	该目录中存放的是该项目的源代码	
gen	该目录下的文件全部由 ADT 自动生成的，不需要去修改。通常该目录下有 R.java 文件和 AIDL 定义生产的相关服务 stub 文件。最常见的是 R.java 文件，该文件相当于项目的字典，为项目中用户界面、字符串、图片等资源都会在该类中创建其唯一的 ID，当项目中使用这些资源时，会通过该 ID 得到资源的引用	
Android x.x.x	该目录中存放项目的 Android 对应版本 jar 包，同时其中还包含项目打包时需要的 META-INF 目录	
Android Dependencies	这是 ADT 的第三方库新的引用方式。当你需要引用第三方库时，只需在项目中新建一个名为 libs 的文件夹，然后将所有第三方包复制到该目录下。ADT 就会自动帮你完成库的引用，Android Dependencies 会自动增加相应的对 jar 包的引用	
assets	资源路径，不会在 R 文件注册。该目录用于存放项目相关的资源文件，在程序中可以使用"getResources.getAssets().open(资源文件名)"得到资源文件的输入流 InputStream 对象，然后可以读该文件内容	
bin	二进制文件，编译好的目标文件。包括 class、资源文件、dex、apk 等	
res	drawable	drawable 开头的三个文件夹用于存储 *.png，*.jpg 等图片资源
	anim	该目录下存放 XML 文件编译为帧序列动画或者自动动画对象
	layout	文件夹存放的是应用程序的界面 GUI 布局文件
	raw	用于存放应用程序所用到的声音等资源。raw 中的文件会被映射到 R.java 文件中，访问的时候直接使用资源 ID 即 R.id.filename；相比较 assets 文件夹下的文件不会被映射到 R.java 中，访问的时候需要 AssetManager 类
	values	存放的是所有 xml 格式的资源描述文件，例如，字符串资源的描述文件 strings.xml、样式的描述文件 styles.xml、颜色描述文件 colors.xml、dimens.xml 尺寸描述文件以及数组描述文件 arrays.xml 等
AndroidManifest.xml	清单文件，用程序的配置与管理，设置相关权限，该文件在软件安装的时候被读取。如：Android 中的四大组件(Activity、ContentProvider、BroadcastReceiver、Service)都需要在该文件中注册，程序运行所需的权限也需要在此文件中声明，例如，电话、短信、互联网、访问 SD 卡等	

系统程序代码：

```
package com.shnu.helloworld;
public class MainActivity extends ActionBarActivity {
    @Override
    protected void onCreate(Bundle savedInstanceState) {
        super.onCreate(savedInstanceState);
        setContentView(R.layout.activity_main);
    }
}
```

资源文件 string.xml：

```
<?xml version="1.0" encoding="utf-8"?>
<resources>
    <string name="app_name">HelloWorld</string>
    <string name="hello_world">Hello world!</string>
    <string name="action_settings">Settings</string>
</resources>
```

布局文件 Layout：

```
<FrameLayout xmlns:android="http://schemas.android.com/apk/res/android"
    xmlns:tools="http://schemas.android.com/tools"
    android:id="@+id/container"
    android:layout_width="match_parent"
    android:layout_height="match_parent"
    tools:context="com.shnu.helloworld.MainActivity"
    tools:ignore="MergeRootFrame" >
    <EditText
        android:id="@+id/editText1"
        android:layout_width="match_parent"
        android:layout_height="wrap_content"
        android:ems="10"
        android:text="@string/hello_world" >
        <requestFocus />
    </EditText>

</FrameLayout>
```

系统资源与 Java 代码映射文件 R.java：

```
package com.shnu.helloworld;
public final class R {
    public static final class anim {
        public static final int abc_fade_in = 0x7f040000;
        public static final int abc_fade_out = 0x7f040001;
        public static final int abc_slide_in_bottom = 0x7f040002;
        public static final int abc_slide_in_top = 0x7f040003;
        public static final int abc_slide_out_bottom = 0x7f040004;
```

```
        public static final int abc_slide_out_top = 0x7f040005;
    }

AndroidMinfest.xml
<?xml version = "1.0" encoding = "utf - 8"?>
<manifest xmlns:android = "http://schemas.android.com/apk/res/android"
    package = "com.shnu.helloworld"
    android:versionCode = "1"
    android:versionName = "1.0" >

    <uses - sdk
        android:minSdkVersion = "8"
        android:targetSdkVersion = "19" />

    <application
        android:allowBackup = "true"
        android:icon = "@drawable/ic_launcher"
        android:label = "@string/app_name"
        android:theme = "@style/AppTheme" >
        <activity
            android:name = "com.shnu.helloworld.MainActivity"
            android:label = "@string/app_name" >
            <intent - filter>
                <action android:name = "android.intent.action.MAIN" />
                <category android:name = "android.intent.category.LAUNCHER" />
            </intent - filter>
        </activity>
    </application>
</manifest>
    protected void onRestoreInstanceState(Bundle savedInstanceState) {
        super.onRestoreInstanceState(savedInstanceState);
        ts.showState();
    }
```

这些文件归纳起来可以分三大类，其逻辑关系如图 2-16 所示。

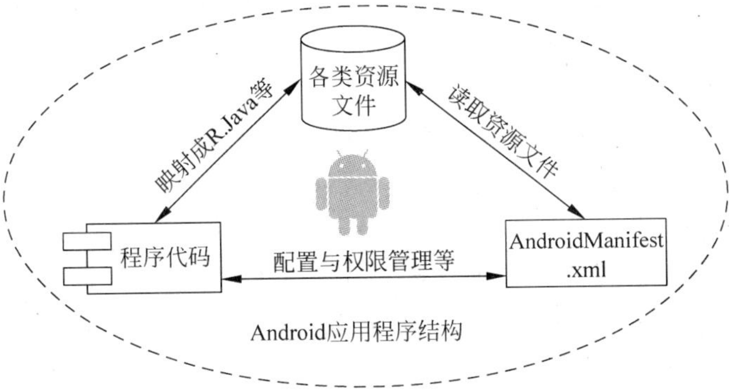

图 2-16　Android 项目源文件结构关系

Android 应用程序做到了(并提倡)将应用程序资源与代码完全分离。Android 资源系统保存所有与代码无关资源的存根。可以使用 Resources 类访问应用程序的资源；与应用程序相关联的资源实例可以通过 Context. getResources()得到。

2.4 Android 应用程序的签名

Android 通过数字签名来标识应用程序的作者和在应用程序之间建立信任关系，这个数字签名由应用程序的开发者来完成，并不需要第三方权威的数字证书签名机构认证，其作用是让应用程序包自我认证和识别。

2.4.1 Android 应用程序使用数字证书的作用

所有 Android 应用程序都必须有数字证书签名，Android 应用程序在安装的时候，Android 操作系统会检查该应用程序有无数字证书签名，没有数字证书签名的应用程序不会安装到系统中。如果要正式发布一个 Android 应用程序，必须使用一个合法的私钥生成的数字证书来给程序签名，Eclipse 的 ADT 插件生成的调试证书不能用来正式发布程序。任何数字证书都是有有效期的，Android 只是在应用程序安装的时候才会检查证书的有效期。如果程序已经安装在系统中，即使证书过期也不会影响程序的正常功能。Android 应用程序使用数字证书的有很多独特的作用，下面分别介绍。

1. Android 程序管理维护

Android 应用程序要求应用所在的包名与数字签名一一对应。当 Android 应用程序的包名相同，且新版程序和旧版程序的数字证书相同时，Android 系统才会认为这两个程序是同一个程序的不同版本。如果新版程序和旧版程序的数字证书不相同，则 Android 系统认为它们是不同的程序，并产生冲突，会要求新程序更改包名。

如果希望用户无缝升级到新的版本，那么必须用同一个证书进行签名。这是由于只有以同一个证书签名，系统才会允许安装升级的应用程序。如果采用了不同的证书，那么系统会要求应用程序采用不同的包名称，在这种情况下相当于安装了一个全新的应用程序。如果想升级应用程序，签名证书要相同，包名称也要相同!

2. 应用程序模块化

Android 系统允许拥有同一个数字签名的程序运行在一个进程中，Android 程序会将它们视为同一个程序。所以开发者可以将自己的程序分模块开发，把应用程序以模块的方式进行部署，而用户可以独立地升级其中的一个模块。

3. 代码或者数据共享

Android 提供了基于数字证书的权限赋予机制，应用程序可以和其他的程序共享该功能或者数据给那些与自己拥有相同数字证书的程序。一个应用程序就可以为另一个以相同证书签名的应用程序公开自己的功能。以同一个证书对多个应用程序进行签名，利用基于签名的权限检查，就可以在应用程序间以安全的方式共享代码和数据了。

2.4.2 Android 应用程序数字证书的使用

1. 数字证书的获得

1) debug 数字证书

为了方便开发者使用，Eclipse＋ADT＋Android SDK 在首次使用时，ADT 会自动在用户目录中(通常是/Documents and Settings/用户名/.android/)建立一个 debug.keystore 数字证书库文件,这里面保存着 debug 数字证书,这个数字证书的有效期通常是自首次创建后密钥后一年时间,超过一年该数字证书将失效。数字证书失效后,开发者可以删除 debug.keystore 数字证书库文件,然后再次运行 Eclipse+ADT+Android SDK 集成开发环境,系统又会自动为我们创建一个新的密钥库文件,又可以再使用一年。在使用集成开发环境时候,会默认使用这个密钥对 Android 应用程序进行数字签名。

2) 开发者使用 Java JDK 中的 Keytool 工具制作数字证书

Keytool 是一个 Java 数据证书的管理工具,Keytool 将密钥(key)和证书(certificates)存在一个称为 keystore 的文件中。

在 keystore 中包含两种数据:

第一,密钥实体(Key entity)。密钥(secret key)是私钥和配对公钥(采用非对称加密)。

第二,可信任的证书实体(trusted certificate entries),其中只包含公钥。

Keytool 工具使用方法非常简单。命令如下:

```
keytool -genkey -v -keystore shnu_client.keystore -alias shnu_client -keyalg RSA -
validity 20000
```

如图 2-17 所示。

```
C:\Java\jdk1.8.0\bin>keytool -genkey -v -keystore shnu_client.keystore -alias sh
nu_client -keyalg RSA -validity 20000
输入密钥库口令:
再次输入新口令:
您的名字与姓氏是什么?
  [Unknown]:  Li
您的组织单位名称是什么?
  [Unknown]:  Computer Department
您的组织名称是什么?
  [Unknown]:  Shanghai Normal University
您所在的城市或区域名称是什么?
  [Unknown]:  Shanghai
您所在的省/市/自治区名称是什么?
  [Unknown]:  Shanghai
该单位的双字母国家/地区代码是什么?
  [Unknown]:  China
CN=Li, OU=Computer Department, O=Shanghai Normal University, L=Shanghai, ST=Shan
ghai, C=China是否正确?
  [否]:  是

正在为以下对象生成 2,048 位RSA密钥对和自签名证书 (SHA256withRSA) (有效期为 20,00
0 天):
         CN=Li, OU=Computer Department, O=Shanghai Normal University, L=Shanghai
, ST=Shanghai, C=China
输入 <shnu_client> 的密钥口令
        (如果和密钥库口令相同, 按回车):
[正在存储shnu_client.keystore]

C:\Java\jdk1.8.0\bin>
```

图 2-17 Keytool 工具的使用

如图 2-17 所示的命令执行完毕后,会在当前目录下生成 shnu_client.keystore 文件。

通常情况下,数字证书的有效期要远远大于 Android 应用的软件使用生命周期(至少在 20 年以上,如,Google play 强制要求所有应用程序数字证书的有效期要持续到 2033 年 10 月 22 日以后)。否则一旦数字证书失效,持有该数字证书的程序将不能正常升级。

2. Android 应用程序的数字证书签名

当要一个 Android 应用程序要正式发布程序时，开发者就必须数字证书给 apk 包签名。签名的方法有两种：

(1) 在命令行下使用 JDK 中的 Jarsigner(用于使用数字证书签名)来给 apk 包签名。

(2) 使用 ADT Export Wizard 进行签名(如果没有数字证书，可能需要生成数字证书)。

这里仅介绍使用 ADT Export Wizard 进行签名的方法。

在 Eclipse 中，右击应用程序工程，在弹出的快捷菜单选择 Android Tools→Export Signed Application Package 选项，如图 2-18 所示。

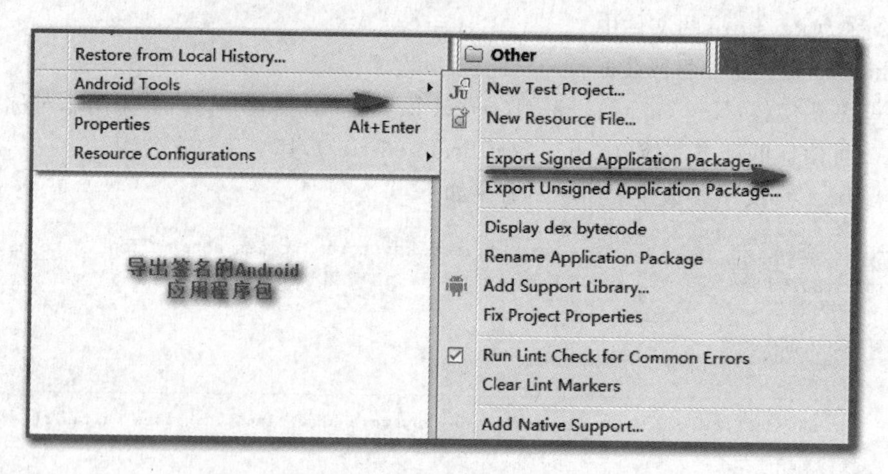

图 2-18　Eclipse 中对项目进行数字签名及导出

选择密钥库(如上面生成的 shnu_client. keystore)，并输入密钥库密码，如图 2-19 所示。

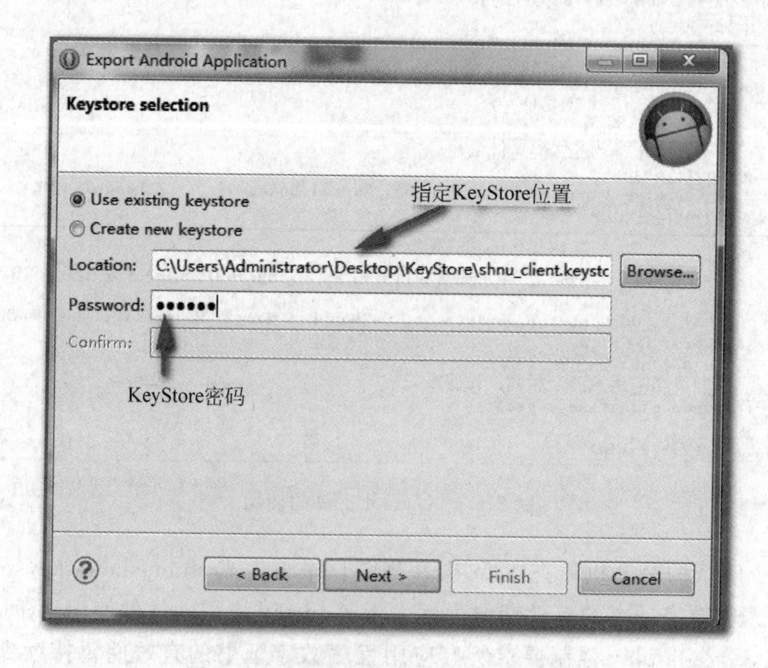

图 2-19　选择密钥库的位置并输入密码

选择证书的存放路径,并输入密钥库密码,如图 2-20 所示。

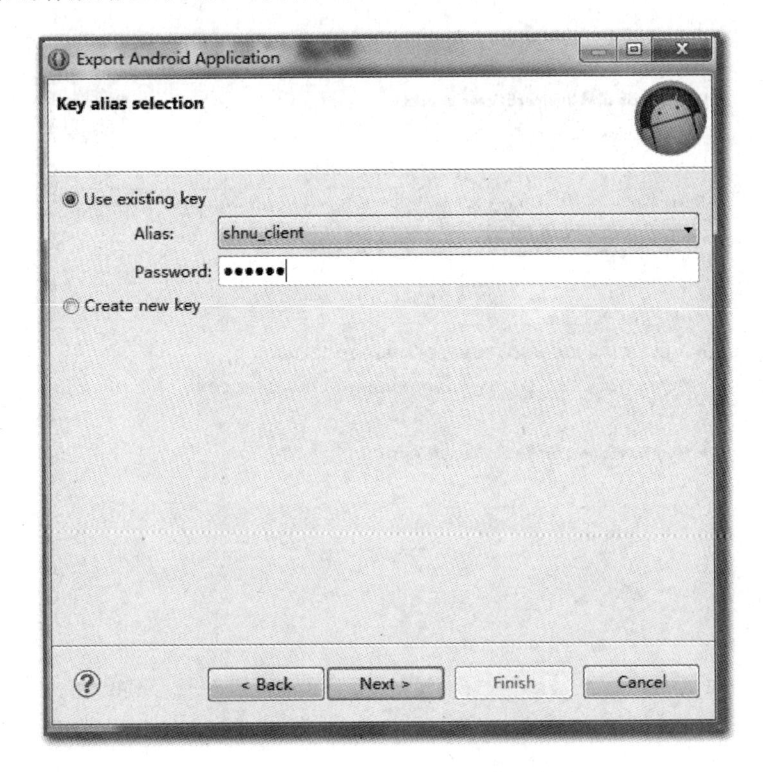

图 2-20　选择证书的位置并输入密码

选择数字签名后要生成 Android 应用程序 apk 文件名,如图 2-21 所示。

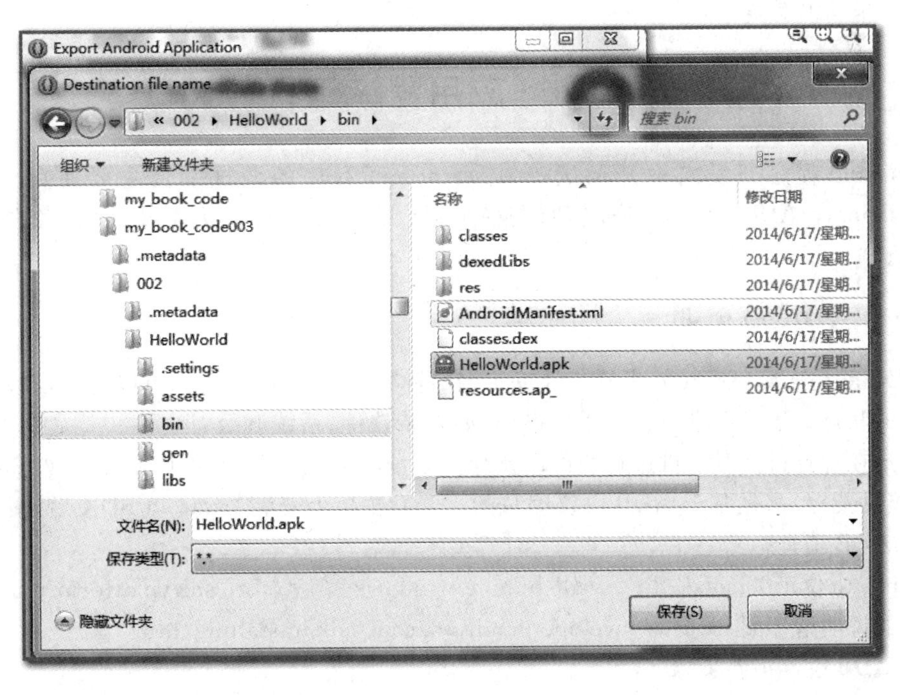

图 2-21　选择生成数字签名的 apk 文件名

数字签名完毕，显示签名后的概要信息，如图 2-22 所示。

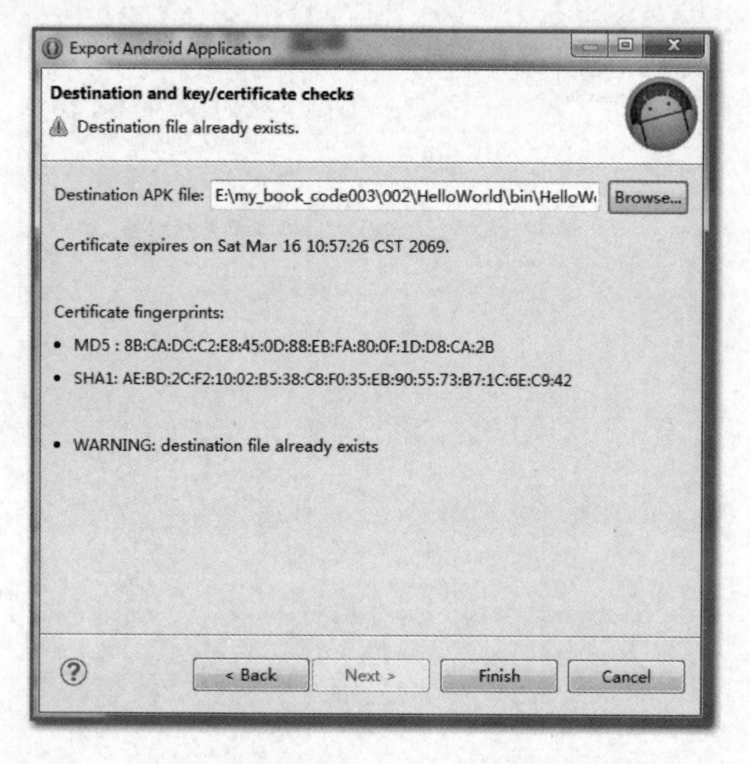

图 2-22 数字签名完成后显示的概要信息

至此，就完成了应用程序的编译、数字签名、打包工作。签名后的 apk 应用程序可以在 Android 市场上发布。

2.5 Android 应用程序运行与调试

Android 程序开发离不开调试工具。Android 程序的调试工具主要使用 Android Debug Bridge (ADB)协议，来完成 PC 与模拟器或真实手机或平板 Android 设备之间的调试通信。

2.5.1 ADB 的使用

ADB 就是连接 Android 手机与计算机端的桥梁，借助 ADB 工具，可以管理设备或手机模拟器的状态。还可以进行很多手机操作，如安装软件、系统升级、运行 shell 命令等等，可以让用户在计算机上对手机进行全面的操作。目前所有的手机助手类的软件(如 360 手机助手、豌豆荚等)都是基于 ADB 协议编写的。开发主机上安装 Android SDK 后，便已经安装好了 ADB 工具。

adb. exe 位于＜install-dir＞\adt-bundle-windows-＜版本＞\sdk\platform-tools，有关这方面的说明请参阅 http://developer. android. com/sdk/installing. html。

1. ADB 设置主要步骤

(1) 在 Android Manifest 中将应用程序声明为“可调试”。在使用 Eclipse 时，可跳过该步骤，因为直接从 Eclipse IDE 运行应用程序时会自动启用调试。在 AndroidManifest. xml 文件中将 android:debuggable="true"添加至＜application＞元素。

(2) 打开手机或平板电脑设备上的"USB 调试"模式。通常在设备上，设置方法为转至"设置"→"应用程序"→"开发"并启用"USB 调试"(在 Android 4.0 设备上，设置路径为"设置"→"开发人员选项")。ADB 会自动与本机模拟器连接，无须设置。

(3) 设置系统以检测设备。如果在 Windows 上开发，则需要安装 ADB 的 USB 驱动程序。有关安装指南和原始设备制造商驱动程序的链接，请参阅 OEM USB Drivers 文档。

2. ADB 的启动过程

(1) 在启动 ADB 客户端时，客户端会先检查是否已经有 ADB 服务器进程正在运行。如果没有，则会启动服务器进程。当服务器启动时，它会绑定至本地 TCP 端口 5037 并监听从 ADB 客户端发出的命令，所有 ADB 客户端都使用端口 5037 与 ADB 服务器通信。

(2) 然后，服务器会建立与所有正在运行的模拟器/设备实例的连接。它会通过在范围 5555～5585(模拟器/设备使用的范围)中扫描奇数编号的端口找到模拟器/设备实例。在找到 ADB 守护程序后，建立与该端口之间的连接。

(3) 每个模拟器/设备实例获取一对序列端口——一个偶数编号的端口用于控制台连接，一个奇数编号的端口用于 ADB 连接。例如，

```
Emulator 1, console: 5554 Emulator 1, adb: 5555
```

一旦服务器建立与模拟器或设备实例之间的连接，就可使用 ADB 命令来控制和访问那些实例。由于服务器会管理指向模拟器/设备实例的连接并处理来自多个 ADB 客户端的命令，可通过任何客户端(或脚本)控制任何模拟器/设备实例。

3. ADB 命令

表 2-3 中的命令可帮助将接受调试的应用程序从命令行转移至目标设备或模拟设备。这一点非常有用，尤其是在没有 ssh 终端连接可用时。

表 2-3　ADB 主要命令格式

ADB 命令	说　　明
adb push ＜local＞ ＜remote＞	copy file/dir to device
adb pull ＜remote＞ [＜local＞]	copy file/dir from device
adb sync [＜directory＞]	copy host－＞device only if changed
adb shell	run remote shell interactively
adb shell ＜command＞	run remote shell command＜
adb emu ＜command＞	- run emulator console command＜
adb logcat [＜filter-spec＞]	View device log
adb forward ＜local＞ ＜remote＞	forward socket connections
adb jdwp	list PIDs of processes hosting a JDWP transport
adb install [-l] [-r] [-s] ＜file＞	push this package file to the device and install it
adb uninstall [-k] ＜package＞	remove this app package from device

详细信息请参阅 http://developer.android.com/guide/developing/tools/adb.html。

4. 常见的 ADB 命令格式

可以在开发机上的命令行或脚本上发送 Android 命令，使用方法：

```
adb [ -d| -e| -s <serialNumber>] <command>
```

当发出一个命令，系统启用 Android 客户端。客户端并不与模拟器实例关联，所以，如果双服务器/设备是运行中的，需要用-d 选项去为应用选择被控制的目标实例。关于使用这个选项的更多信息，可以查看模拟器/设备实例术语控制命令。

5. ADB 查询模拟器/设备实例

在发布 ADB 命令之前，有必要知道什么样的模拟器/设备实例与 ADB 服务器是相连的。可以通过使用 devices 命令来得到一系列相关联的模拟器/设备：

```
adb devices
D:\adt-bundle-windows-x86-20140321\sdk\platform-tools\adb devices
    List of devices attached
    ZTEU930HD   device
    emulator-5554   device
```

如果当前没有模拟器/设备运行，则 ADB 返回 no device。

6. 给特定的模拟器/设备实例发送命令

如果有多个模拟器/设备实例在运行，在发布 ADB 命令时需要指定一个目标实例。这样做，请使用-s 选项的命令。命令格式是：

```
adb -s <serialNumber> <command>
```

如上所示，给一个命令指定了目标实例，这个目标实例使用由 ADB 分配的序列号。可以使用 devices 命令来获得运行着的模拟器/设备实例的序列号。示例如下：

```
adb -s emulator-5556 install helloWorld.apk
```

如果没有指定一个目标模拟器/设备实例就执行“-s”这个命令，ADB 会产生一个错误。

7. 安装软件

使用 ADB 从开发计算机上复制一个应用程序，并且将其安装在一个模拟器/设备实例。install 命令必须指定所要安装的.apk 文件的路径：

```
adb install <path_to_apk>
```

为了获取更多的关于怎样创建一个可以安装在模拟器/设备实例上的.apk 文件的信息，可参阅 Android Asset Packaging Tool（AAPT）。

要注意的是，如果正在使用 Eclipse IDE 并且已经安装过 ADT 插件，那么就不需要直接使用 ADB(或者 AAPT)去安装模拟器/设备上的应用程序。ADT 插件会代你全权处理应用程序的打包和安装。

8. 转发端口

可以使用 forward 命令进行任意端口的转发——一个模拟器/设备实例的某一特定主机端口向另一不同端口的转发请求。下面演示了如何建立从主机端口 6100 到模拟器/设备端口 7100 的转发。

```
adb forward tcp:6100 tcp:7100
```

同样，可以使用 adb 来建立命名为抽象的 UNIX 域套接口，上述过程如下所示：

```
adb forward tcp:6100 local:logd
```

9. 复制文件到模拟器设备

可以使用 push 命令将文件复制到一个模拟器/设备实例的数据文件或是从数据文件中复制。install 命令只将一个.apk 文件复制到一个特定的位置，与其不同的是，push 命令可复制任意的目录和文件到一个模拟器/设备实例的任何位置。

10. 从模拟器或者设备中复制出文件或目录

使用如下命令：

```
adb pull <remote> <local>
```

将文件或目录复制到模拟器或者设备，使用如下命令：

```
adb push <local> <remote>
```

在这些命令中，<local>和<remote>分别指通向自己的开发机(本地)和模拟器/设备实例(远程)上的目标文件/目录的路径。

下面是一个例子：

```
adb push foo.txt /sdcard/foo.txt
```

11. 启动 shell 命令

Adb 提供了 shell 端，通过 shell 端可以在模拟器或设备上运行各种命令。

```
adb -s ZTEHD930 shell
shell@android:/$ ls  /system/bin
```

可以运行 Android 手机中相关 Linux 系统命令。

12. 软件测试

Android SDK 中提供了软件测试工具 Monkey，Monkey 工具采用随机重复的方法测试软件。如可以使用下列命令启动 com.shnu.helloworld 软件并触发 1000 个事件：

```
adb shell monkey *v -p com.shnu.helloworld  1000
```

有关 Monkey 的命令使用方法参见 http://developer.android.com/tools/help/monkey.html。

2.5.2 DDMS 介绍

DDMS 的全称是 Dalvik Debug Monitor Service，DDMS 将搭建起 IDE 与测试终端(Emulator 或者 connected device)的链接，它们应用各自独立的端口监听调试器的信息，DDMS 可以实时监测到测试终端的连接情况。

当有新的测试终端连接后，DDMS 将捕捉到终端的 ID，并通过 adb 建立调试器，从而实现发送指令到测试终端的目的。它可以完成测试设备截屏,查看特定的进程以及堆信息、Logcat、广播状态信息、模拟电话呼叫、接收 SMS、虚拟地理坐标等。启动 DDMS 的方法如下:直接双击 ddms. bat 运行;或者在 Eclipse 中打开 DDMS 视图。

DDMS 监听第一个终端 App 进程的端口为 8600,下一个 App 进程将分配 8601,其他进程将按照这个顺序依次类推。DDMS 通过 8700 端口接收所有终端的指令。在 GUI 的左上角可以看到标签为 Devices 的面板,在这里可以查看到所有与 DDMS 连接的终端的详细信息,以及每个终端正在运行的 App 进程,每个进程最右边相对应的是 PID 与调试器链接的端口。

在面板的右上角有一排很重要的按键,分别是 Debug the selected process、Update Threads、Update Heap、Stop Process 和 ScreenShot(见图 2-23)。

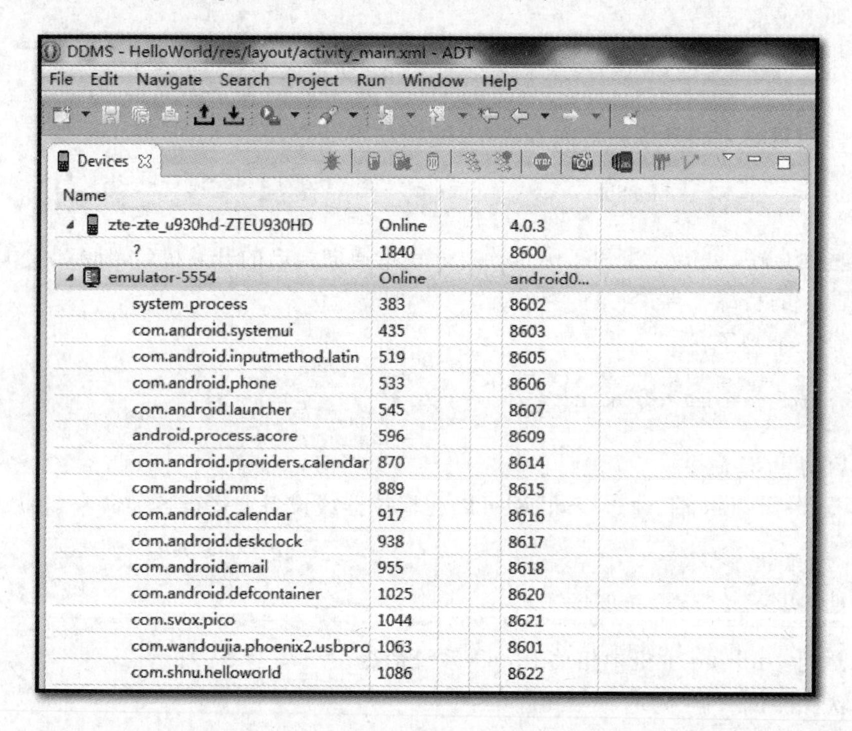

图 2-23 Eclipse+ADT 中 DDMS 调试窗口

DDMS 的使用

DDMS 可以与模拟器通信,模拟接听电话、不同网络情况、发送或接收 SMS 消息、发送虚拟地址坐标用于测试 GPS 功能等。除此之外,DDMS 还可以完成模拟器或外接手机或平板电脑的文件浏览、上传下载等(见图 2-24 和图 2-25)。

如使用 DDMS,可以浏览刚才完成的“HelloWorld”程序,在模拟器、手机或平板电脑中安装在/data/app/com. shnu. helloworld. apk 路径下。/system/app 目录主要存放的是常规下载的应用程序(通常是手机系统内置 App 或厂商预装的 App),可以看到都是以. apk 结尾的文件。在这个文件夹下的程序为系统内置默认的应用组件。如闹钟、Gmail、日历、计算器等应用,每个应用都有 *. apk 和对应的 *. odex 文件成对出现,参见图 2-26。

DDMS 是开发人员最好的调试工具,是 Android 开发的人员必须掌握的调试工具。

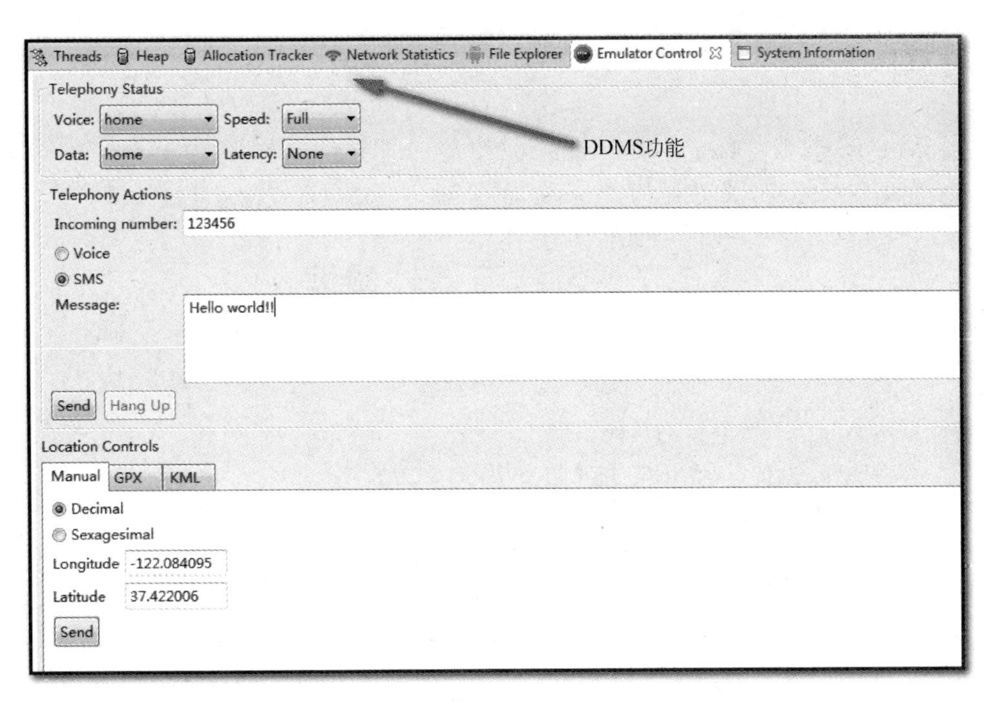

图 2-24　DDMS 对模拟器的控制界面

Threads　Heap　Allocation Tracker　Network Statistics　File Explorer　Emulator Control　System Information

Name	Size	Date	Time	Permissions	Info
acct		2014-04-30	20:58	drwxr-xr-x	
cache		2014-04-30	20:58	drwxrwx---	
config		2014-04-30	20:58	dr-x------	
d		2014-04-30	20:58	lrwxrwxrwx	-> /sys/kernel/debug
data		2014-04-30	00:00	drwxrwx--x	
anr		2014-04-30	00:09	drwxrwxr-x	
app		2014-04-30	21:09	drwxrwx--x	
ApiDemos.apk	4764089	2014-01-17	16:18	-rw-r--r--	com.example.android.apis
ApiDemos.odex	1579120	2014-01-17	16:18	-rw-r--r--	
CubeLiveWallpapers.apk	19329	2014-01-17	16:17	-rw-r--r--	com.example.android.livecubes
CubeLiveWallpapers.odex	15080	2014-01-17	16:17	-rw-r--r--	
GestureBuilder.apk	27684	2014-01-17	16:17	-rw-r--r--	com.android.gesture.builder
GestureBuilder.odex	23240	2014-01-17	16:17	-rw-r--r--	
SmokeTest.apk	7977	2014-01-17	16:15	-rw-r--r--	com.android.smoketest.tests
SmokeTest.odex	12496	2014-01-17	16:15	-rw-r--r--	
SmokeTestApp.apk	3359	2014-01-17	16:15	-rw-r--r--	com.android.smoketest
SmokeTestApp.odex	2008	2014-01-17	16:15	-rw-r--r--	
SoftKeyboard.apk	44060	2014-01-17	16:17	-rw-r--r--	com.example.android.softkeyboar
SoftKeyboard.odex	28128	2014-01-17	16:17	-rw-r--r--	
WidgetPreview.apk	19237	2014-01-17	16:17	-rw-r--r--	com.android.widgetpreview
WidgetPreview.odex	10800	2014-01-17	16:17	-rw-r--r--	
com.shnu.helloworld-1.apk	800189	2014-04-30	21:08	-rw-r--r--	com.shnu.helloworld
com.wandoujia.phoenix2-1.apk	7802188	2014-04-29	23:57	-rw-r--r--	com.wandoujia.phoenix2
com.wandoujia.phoenix2.usbproxy-1.apk	583471	2014-04-29	23:54	-rw-r--r--	com.wandoujia.phoenix2.usbproxy
app-asec		2014-04-29	23:18	drwx------	

Hello World程序安装位置

图 2-25　刚才生成的 HelloWorld 在系统中的存放位置

Saved Filters
All messages (no filters) (3211)
com.shnu.helloworld (Session Filter)

Search for messages. Accepts Java regexes. Prefix with pid:, app:, tag: or text: to limit scope.

L...	Time	PID	TID	Application	Tag	Text
D	04-30 21:09:14.229	1086	1086	com.shnu.helloworld	dalvikvm	Not late-enabling CheckJNI (already on)
D	04-30 21:09:18.959	1086	1086	com.shnu.helloworld	gralloc_goldfish	Emulator without GPU emulation detected
I	04-30 22:44:42.239	1086	1086	com.shnu.helloworld	Choreographer	Skipped 34 frames! The application may be doing too much work on its mai

图 2-26　Logcat 输出信息

2.6 本章小结

通过本章的学习，我们已经掌握 Android 开发环境的搭建、Android 开发工具的使用、Android 应用程序的逻辑组成结构、Android 模拟器的使用以及应用程序的调试。

2.7 习题与课外阅读

2.7.1 习题

(1) 简述 Android 模拟器的功能。

(2) 使用 ADB 上传一个 MP3 文件到模拟器，并播放该文件。

(3) 简述 ADB 协议原理及功能。

2.7.2 课外阅读

(1) 访问下列技术网站，了解一下 Android 应用系统开发过程：

http://developer.android.com/training/index.html。

(2) 下载 SensorSimulator 传感器仿真工具，并仿真加速传感器。

http://code.google.com/p/openintents/wiki/SensorSimulator。

(3) 访问下列网站并了解一下手机云测试服务。

http://www.testin.cn。

第3章 Activity及生命周期

一个 Android 应用程序通常是由四大组件 Activity、Service、Content Provider、Broadcast Receiver 中的组件构成。由于 Android 并不提供 Linux 终端界面应用，Android 应用的用户界面只能由 Activity 和 UI 控件(View)提供，Activity 作为 UI 控件的承载体(或容器)，其生命周期直接影响一个 Android 应用的运行及状态，因此本章重点介绍 Activity 及其生命周期。

本章学习目标：

- 掌握 Activity 的概念及创建方法；
- 掌握 Activity 的生命周期及运行状态；
- 掌握保存和恢复 Activity 运行状态数据的方法。

3.1 Activity 简介

Activity 是 Android 基本组件之一，它与 Service、ContentProvider、BroadcastReceiver 合称为 Android 应用程序的四大核心组件，也号称"四大金刚"。Activity 主要提供用户交互界面 UI 控件(View)的载体，UI 控件是通过布局管理器(Layout Manager)摆放到 Activity 上的。

大多数 Android 应用包含一个或多个 Activity，其中一个 Activity 作为主(main) Activity，通常该 Activity 作为应用程序界面的入口。一个 Activity 在调用其他 Activity 时，被保存在一个 Activity"后进先出"堆栈中。当用户单击返回按钮时，该 Activity 从堆栈中弹出恢复运行。在某种程度上说，一个 Activity 有点像用户用浏览器进行一个网页浏览的前进与后退操作。

一个 Activity 可以展示 Android 系统提供或用户自定义的 View 元素。如一个电话拨号应用程序可以包括一个用于显示联系人的列表的 Activity，以及查看通话记录的 Activity。尽管它们暴露给用户是一个内聚的用户界面与应用，但其中每个 Activity 都与其他的保持独立。每个都是以 Activity 类为父类的子类实现的。

每个 Activity 通常在一个默认的窗口(Window)中渲染。窗口显示的可视内容是由一系列 View 的子类(通常是可见的图形界面控件，如按钮、文本框、复选框等)构成的。

每个 View 均控制着窗口中一块特定的矩形空间。父级 View 包含并组织其子视图的布局。上下文 View 是位于 View 层次根位置的 View 对象。叶节点 View(位于视图层次最底端)在它们控制的矩形中进行绘制，并对用户对其直接操作做出响应。所以，View 是 Activity 与用户进行交互的界面。比如说，View 可以显示一个文本框，用户点击的时候产生动作事件。Android 为用户提供了很多既定的 View，包括按钮、文本域、卷轴、菜单项、复选框等，如图 3-1 所示。

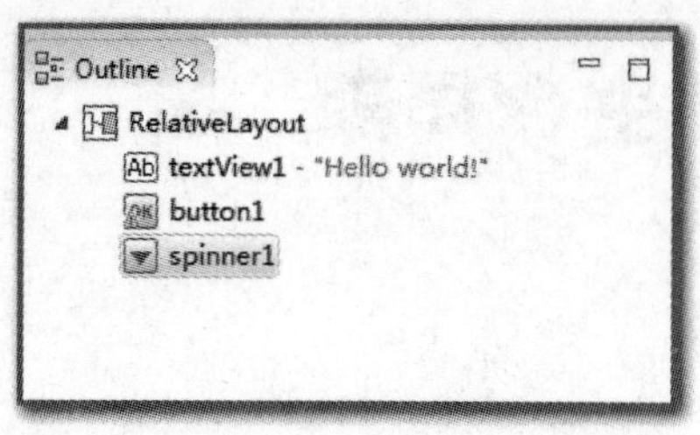

图 3-1　View 的逻辑层次

视图层次是在 layout. xml 等布局文件中定义的，并由 Activity. setContentView()方法，解析并装载在 Activity 的窗口之中渲染。

一个 Activity 实例对应于一个应用程序窗口，它本身不提供用户界面元素，只是提供所有 UI 控件(View)摆放和依附的容器，每一个 Activity 都有至少一种布局管理器来管理依附在其上的 View 控件。Activity 可以获得焦点，接收用户事件，并把事件分发给依附在 Activity 之上的控件进行处理，完成应用程序的人机交互。

3.2　Activity 生命周期

每一个 Activity 都有生命周期，在不同的周期阶段运行不同的函数，这些状态对于一个 Android 应用状态至关重要。

打个比方，Activity 相当于“舞台”，布局管理器相当于“导演”，UI 控件(如文本框、按钮等)相当于“演员”。“演员”(UI 控件)不能自行上“舞台”(Activity)，而只能根据“导演”布局管理器(Layout Manager)决定其在舞台的位置。

“舞台”Activity 的生命周期，对一部电影或戏至关重要，因为假如“舞台”Activity 都不见了，怎么还能谈得上“演戏”，控件都无法显示，也就无法与用户交互了。

由于 Android 应用程序运行环境相对比较复杂，“舞台”Activity 可能随时会不可见或消失，如 Android 程序运行时有可能有电话呼入、收到短信、网络中断、系统内存不足、系统电量不足等等复杂状况或场景，这些状况对 Android 的 Activity 运行状态有很大影响。我们了解与掌握 Activity 的生命周期的目的在于：掌握在 Activity 不同生命周期所运行对应的函数，在相应的函数内，调用开发的逻辑代码，来完成系统所需要的功能。

例如，我们做了一个基于 Android 手机社交软件，在一个 Activity 发送信息界面中，我们准备给朋友发送信息，刚写完信息，正欲发送的时候，突然来了一个电话呼入，系统界面完全跳到电话应用 Activity 界面，等通话结束返回到我们的 Activity 发送消息界面，我们希望我们原先写的消息还存在，可以继续编写或发送消息。这种情况下，我们必须要深入了解 Activity 的生命周期，在运行状态发生变化时应在相应的函数中保存、恢复数据。

Activity 生命周期中的状态转换说明：Activity 拥有四种基本状态，如图 3-2 所示。

(1) Actived/Runing：一个新的 Activity 入栈后，它在屏幕最前端，处于栈的最顶端，处于可见并且可交互的激活状态。

(2) Paused：当 Activity 被另一个透明或者 Dialog 样式的 Activity 覆盖时的状态。此时它依然与窗口管理器保持连接，系统继续维护其内部状态，所以它仍然可见，但它已经失去了焦点，故不可与用户交互。

(3) Stoped：当 Activity 被另外一个 Activity 覆盖、失去焦点并不可见时的状态。

(4) Killed：Activity 被系统杀死回收或者没有被启动时的状态。

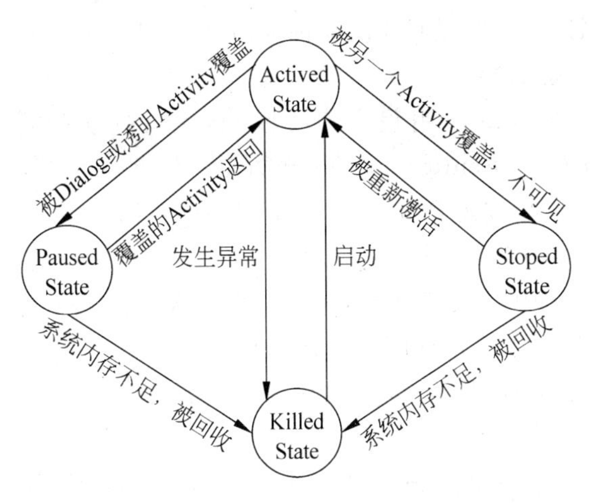

图 3-2　Activity 的状态转换

当一个 Activity 实例被创建、销毁或者启动另外一个 Activity 时，它在这四种状态之间进行转换，这种转换的发生依赖于用户程序的动作。表 3-1 说明了 Activity 在不同状态间转换的时机和条件。

表 3-1　Activity 生命周期与相关方法

Activity 状态转换逻辑图	方法名称	描　　述
Activity starts onCreate() onStart() onResume() Activity is running Another activity comes in front of the activity onPause() The activity is no longer visible onStop() onDestroy() Activity is shut down User navigates back to the activity Process is killed Other applications need memory The activity comes to the foreground onRestart() The activity comes to the foreground	onCreate()	在 Activity 首次被创建的时调用。是所有初始化设置的地方，常用来创建视图、绑定数据至列表等，或恢复曾经有状态记录。总继之以 onStart()
	onStart()	当 Activity 正要变得为用户所见视图时被调用。 当 Activity 转向前台时继之以 onResume()； 当 Activity 变为隐藏时继之以 onStop()
	onRestart()	在 Activity 停止后，在再次启动之前被调用。总继之以 onStart()
	onResume()	在 Activity 开始与用户进行交互之前被调用。此时 Activity 位于堆栈顶部，并接受用户输入。继之以 onPause()
	onPause()	当系统将要启动另一个 Activity 时调用。此方法主要用来将未保存状态数据，停止耗费 CPU 的动作等。 当 Activity 重新回到前台是继以 onResume()； 当 Activity 变为用户不可见时继以 onStop()
	onStop()	当 Activity 不再为用户可见时调用此方法。如果 Activity 再次回到前台跟用户交互，则继以 onRestart()；如果关闭 Activity，则继以 onDestroy()
	onDestroy()	在 Activity 销毁前调用。这是 Activity 接收的最后一个调用。这可能是调用 finish()方法或者因为系统需要空间所以临时的销毁了此 Activity 的实例时。可用 isFinishing()方法来区分这两种情况

Activity 在生命周期不同阶段对应的运行函数，如表 3-1 所示。

Activity 常见运行状态说明：

(1) 当打开一个 Android 应用程序的时候，会先后依次执行该程序主 Activity 的 onCreate()－＞onStart()－＞onResume()三个方法。

(2) 当退出程序(或者按返回键)，会先后依次执行该 Activity 的 onPause()－＞onStop()－＞onDestroy()三个方法。

(3) 打开应用程序运行时，突然有电话呼入，Activity 会先后依次执行了 onPause()－＞onStop()这两个方法，这时候应用程序并没有销毁。

(4) 而当再次启动该 Android 应用程序时，则会先后依次执行了 onRestart()－＞onStart()－＞onResume()三个方法。

(5) 当按 HOME 键，然后再次进入 Activity 应用时，该应用的状态应该是和按 HOME 键之前的状态是一样的。所以，大多数情况下可以在 onPause()里面保存一些数据和状态。而在 onResume()里面来恢复数据。

3.3 Activity 生命周期教学案例

在总体上了解一个 Activity 生命周期后，我们用一个实验来验证一下 Activity 的生命周期。接下来设计一个教学案例，让程序自动打印出 Activity 的生命周期不同阶段所运行的方法，来获取第一手验证性资料，这种教学和学习方法，往往比灌输型知识教学效果好得多。

【例 3-1】 教学案例设计：设计一个函数调用顺序历史记录工具，同时在 Toast 和 System. out 打印出来。

"工欲善其事，必先利其器。"首先介绍一个小工具类 ToastLogTool 。ToastLogTool 类主要功能是记录每个调用它的函数，并在上下文 Context 中使用 ToastLogTool. showState()函数，可以通过 Toast、System. out 打印出当前调用函数历史信息。该小工具的编写与使用非常简单，在今后的调试程序案例中我们会经常使用该小工具。下面给出源码：

```
package com.shnu.tools;
import android.content.Context;
import android.widget.Toast;
public class ToastLogTool {
    private int i = 1;
    private String state;
    private Context context;
    public ToastLogTool(Context context, String tag) {
        this.context = context;
        state = tag + "-->";
    }
    public void showState(){
//获取使用 showState()函数的上一级函数名称
String s = new Throwable().getStackTrace()[2].getMethodName();
//拼接历史状态字符串,并显示
```

```
        state = state + (i++) + "." + s + "-->";
        Toast.makeText(context, state, Toast.LENGTH_LONG).show();
        System.out.println(state);
    }
}
```

【例 3-2】 教学案例设计：设计一个 Activity，打印出 Activity 生命周期的每个函数运行状态。并测试在 Activity 运行时，接收到短信后 Activity 的状态；旋转屏幕后 Activity 的状态；单击返回按钮时 Activity 的状态。验证一下 3.1 节有关 Activity 生命周期的内容。

Activity 生命周期的教学案例比较简单，只需派生出一个 Activity 子类 ActivityLifeTime，然后覆盖一下几个有关 Activity 生命周期的重要函数，创建一个 ToastLogTool 工具，在每个 Activity 生命周期函数中，调用一下 showState()函数即可记录 Activity 的状态转换历史记录。

ActivityLifeTime.java 类的内容如下：

```
package com.shnu.activitylifetimedemo;
import android.os.Bundle;
import com.shnu.tools.ToastLogTool;
import android.app.Activity;
public class ActivityLifeTime extends Activity {
    private ToastLogTool ts = new ToastLogTool(this, this.getClass().getName());
    protected void onCreate(Bundle savedInstanceState) {
        super.onCreate(savedInstanceState);
        setContentView(R.layout.activity_life_time);
        ts.showState();
    }
    protected void onDestroy() {
        super.onDestroy();
        ts.showState();
    }
    protected void onPause() {
        super.onPause();
        ts.showState();
    }
    protected void onRestart() {
        super.onRestart();
        ts.showState();
    }
    protected void onRestoreInstanceState(Bundle savedInstanceState) {
        super.onRestoreInstanceState(savedInstanceState);
        ts.showState();
    }
    protected void onResume() {
        super.onResume();
        ts.showState();
    }
    protected void onSaveInstanceState(Bundle outState) {
        super.onSaveInstanceState(outState);
```

```
            ts.showState();
        }
        protected void onStart() {
            super.onStart();
            ts.showState();
        }
        protected void onStop() {
            super.onStop();
            ts.showState();
        }
        public void recreate() {
            super.recreate();
            ts.showState();
        }
        public void finish() {
            super.finish();
            ts.showState();
        }
        public void onBackPressed() {
            super.onBackPressed();
            ts.showState();
        }
            public boolean onSearchRequested() {
            ts.showState();
            return super.onSearchRequested();
        }
}
```

- ActivityLifeTime 的屏幕布局文件 Layout.xml（UI 文件，可以不必手工编写，直接在 Android Eclipse＋ADT 集成开发环境中，以可视化方法来生成相应的.xml 文件）如下：

```
<RelativeLayout xmlns:android = "http://schemas.android.com/apk/res/android"
    xmlns:tools = "http://schemas.android.com/tools"
    android:layout_width = "match_parent"
    android:layout_height = "match_parent"
    android:paddingBottom = "@dimen/activity_vertical_margin"
    android:paddingLeft = "@dimen/activity_horizontal_margin"
    android:paddingRight = "@dimen/activity_horizontal_margin"
    android:paddingTop = "@dimen/activity_vertical_margin"
    tools:context = ".ActivityLifeTime" >

    <TextView
        android:layout_width = "wrap_content"
        android:layout_height = "wrap_content"
        android:text = "@string/hello_world" />

</RelativeLayout>
```

• 资源文件 string. xml 内容如下(可以不必手工编写,直接在 Android Eclipse+ADT 集成开发环境中,以可视化方法来生成相应的. xml 文件):

```
<?xml version = "1.0" encoding = "utf - 8"?>
< resources >
    < string name = "app_name"> ActivityLifeTimeDemo </string >
    < string name = "action_settings"> Settings </string >
    < string name = "hello_world"> Hello world! </string >
</resources >
```

• AndroidMinifest. xml 文件内容如下(可以不必手工编写,直接在 Android Eclipse+ADT 集成开发环境中,以可视化方法来生成相应的 xml 文件):

```
<?xml version = "1.0" encoding = "utf - 8"?>
< manifest xmlns:android = "http://schemas.android.com/apk/res/android"
    package = "com.shnu.activitylifetimedemo"
    android:versionCode = "1"
    android:versionName = "1.0" >

    < uses - sdk
        android:minSdkVersion = "8"
        android:targetSdkVersion = "17" />

    < application
        android:allowBackup = "true"
        android:icon = "@drawable/ic_launcher"
        android:label = "@string/app_name"
        android:theme = "@style/AppTheme" >
        < activity
            android:name = "com.shnu.activitylifetimedemo.ActivityLifeTime"
            android:label = "@string/app_name" >
            < intent - filter >
                < action android:name = "android.intent.action.MAIN" />

                < category android:name = "android.intent.category.LAUNCHER" />
            </intent - filter >
        </activity>
    </application >
</manifest >
```

软件运行后会自动打印出系统调用函数的顺序状态(建议读者做实验尝试,取得第一手资料):

```
ActivityLifeTime -- > 1.performCreate -- > 2.callActivityOnResume -- > 3.onBackPressed -- >
4.performPause -- > 5.callActivityOnStop -- >
```

下面测试电话呼入对 Activity 的生命周期影响。

运行 ActivityLifeTime,打开 DDMS 模仿电话呼叫模拟器,如图 3-3 所示。

可以看到 ActivityLifeTime 运行状态变化如图 3-4 所示。

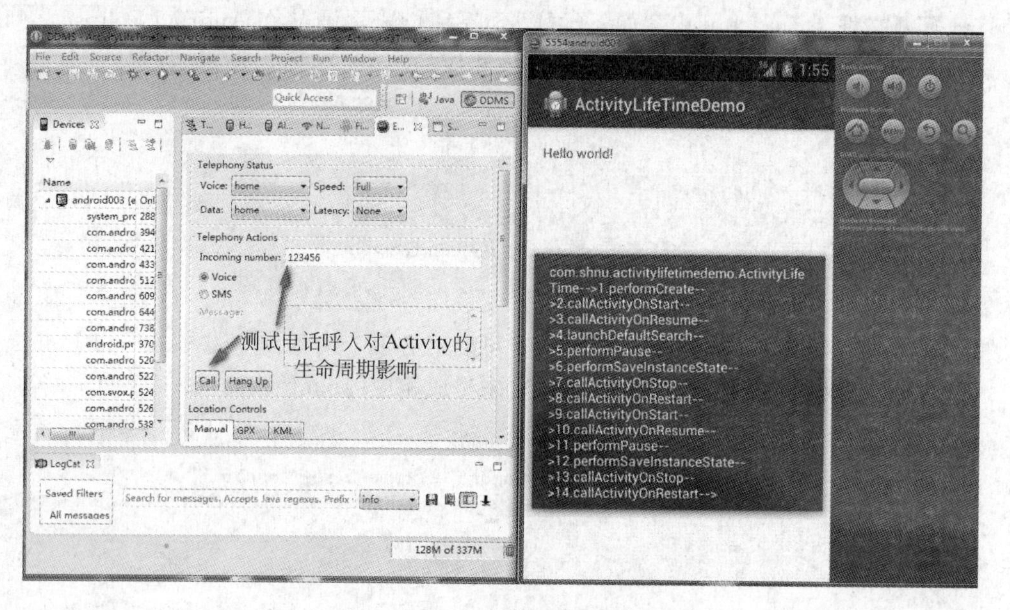

图 3-3　Activity 的生命周期状态转换

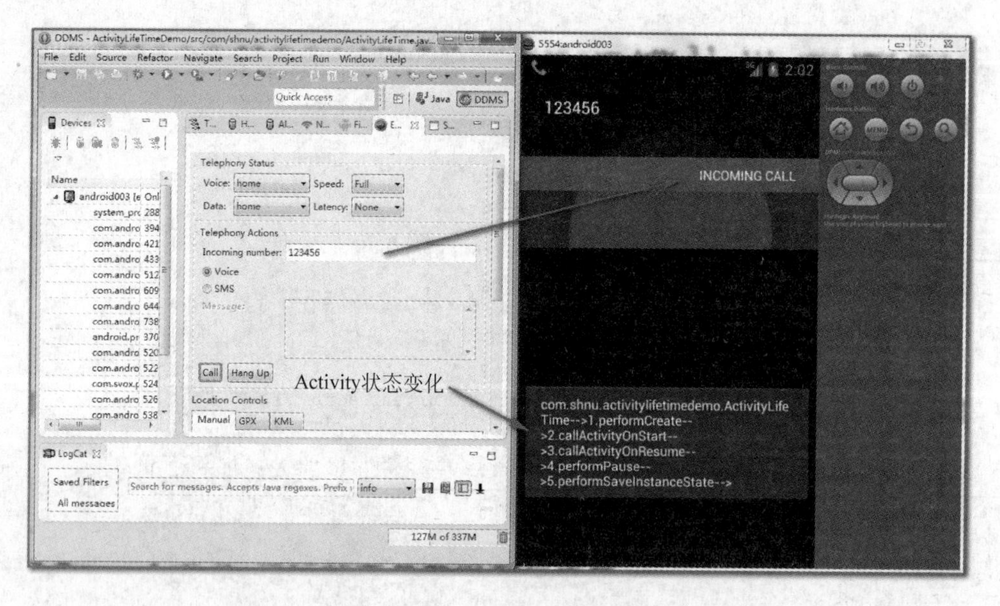

图 3-4　有电话呼入时 Activity 的生命周期状态转换

另外可用 DDMS 模拟发送短信给模拟器，或触摸 Home、返回、查找等按钮，测试这些事件对 ActivityLifeTime 的运行状态。

3.4　Activity 运行状态参数保存与恢复

所谓 Activity 运行状态参数的保存与恢复，大多数情况是指 Activity 在运行过程中，由于 Android 系统其他事件影响，导致 Activity 运行状态改变，Activity 可能被系统压入 Activity 栈中，或因系统内存不足直接被系统摧毁，在这种情况下，原先 Activity 界面中，用户输入的数据可能丢失，一个良好的程序应该能保存恢复这些数据。

在掌握了 Activity 的生命周期以后，让我们来学习 Activity 运行状态参数的保存与恢复。以下几点必须了解：

- 通常情况，调用 onPause()和 onStop()方法后的 Activity 实例仍然存在于内存中，Activity 所持有的信息和状态数据(如用户界面输入的数据)不会消失，当 Activity 重新返回到前台之后，所有的 Activity 运行状态数据(如用户界面输入的数据)仍然会存在。
- 但是当系统内存不足时，调用 onPause()和 onStop()方法后的 activity 在内存中可能会被系统回收。如果这个 activity 再次回到前台，该 Activity 之前的状态数据会丢失，如用户界面中输入的数据文字都消失了(因为这个 Activity 是系统重建立启动的)。
- 为了解决该问题，Android 系统中的 Activity 的 onSaveInstanceState()方法的默认实现会自动保存 Activity 中的某些状态数据，例如 Activity 中各种 UI 控件的状态. Android 应用框架中定义的几乎所有 UI 控件都恰当地实现了 onSaveInstanceState()方法，因此当 activity 被摧毁和重建时，这些 UI 控件会自动保存和恢复状态数据。例如，Activity 运行过程中，EditText 控件会自动保存和恢复控件的属性数据 text。
- 为了解决该问题，可以覆写 Activity 的 onSaveInstanceState()方法，将 Activity 运行状态数据存储到 Bundle 对象中，Bundle 是一个类似 Map 的对象，数据按照(key, value)形式存储，这样即使 Activity 被系统回收或摧毁，当用户重新启动这个 Activity 时，Activity 运行 onCreate()方法时，可以获得该 Bundle 对象，只需从该 Bundle 对象中取出保存的数据，利用这些数据将 Activity 恢复到被回收之前的状态。
- 需要注意的是，onSaveInstanceState()方法并不是一定会被调用的，如果调用系统调用了 onSaveInstanceState()方法，调用将发生在 onPause()或 onStop()方法之前。用户按下返回键退出 Activity 时，或要关闭这个 Activity 时，onSaveInstanceState()方法不会被调用，也就没有必要保存数据以供下次恢复的。

【例 3-3】 教学案例设计：在例 3-2 的 Activity 中增加一个 EditText 控件，系统运行时，显示当前时间，并将该值保存到 Bundle 中，在 Activity 被 destroy()后，再次恢复时候，能显示原先的时间。

该例子在例 3-2 基础上修改得到，这里只列出修改后的部分代码。

```
package com.shnu.activitylifetimedemo;
import java.util.Date;
import android.os.Bundle;
import android.widget.EditText;
import com.shnu.tools.ToastLogTool;
import android.app.Activity;
public class ActivityLifeTime extends Activity {
    private ToastLogTool ts = new ToastLogTool(this, this.getClass().getName());
    EditText ed;
```

```
        protected void onCreate(Bundle savedInstanceState) {
            super.onCreate(savedInstanceState);
            setContentView(R.layout.activity_life_time);
            ts.showState();
            ed = (EditText) this.findViewById(R.id.editText1);
//从 savedInstanceState 中恢复数据，如果没有数据需要恢复 savedInstanceState 为 null
            if (savedInstanceState != null) {
                    ed.setText(savedInstanceState.getString("date"));
                        } else{
            ed.setText(new Date().toString());
        }
    }
protected void onSaveInstanceState(Bundle outState) {

            super.onSaveInstanceState(outState);
// 将数据保存到 outState 对象中，该对象会在重建 activity 时传递给 onCreate 方法
            outState.putString("date", ed.getText().toString());
            ts.showState();
    }
…
}
```

- ActivityLifeTime 的屏幕布局文件 Layout.xml(UI 文件，可以不必手工编写，直接在 Android Eclipse＋ADT 集成开发环境中，以可视化方法来生成相应的.xml 文件)如下：

```
<RelativeLayout xmlns:android = "http://schemas.android.com/apk/res/android"
    xmlns:tools = "http://schemas.android.com/tools"
    android:layout_width = "match_parent"
    android:layout_height = "match_parent"
    android:paddingBottom = "@dimen/activity_vertical_margin"
    android:paddingLeft = "@dimen/activity_horizontal_margin"
    android:paddingRight = "@dimen/activity_horizontal_margin"
    android:paddingTop = "@dimen/activity_vertical_margin"
    tools:context = ".ActivityLifeTime" >

    <TextView
        android:id = "@ + id/textView1"
        android:layout_width = "wrap_content"
        android:layout_height = "wrap_content"
        android:text = "@string/hello_world" />

    <EditText
        android:id = "@ + id/editText1"
        android:layout_width = "wrap_content"
        android:layout_height = "wrap_content"
        android:layout_alignLeft = "@ + id/textView1"
        android:layout_below = "@ + id/textView1"
```

```
        android:layout_marginLeft = "24dp"
        android:layout_marginTop = "54dp"
        android:ems = "10" >
        < requestFocus />
    </EditText >
</RelativeLayout >
```

3.5 本章小结

通过本章的学习，我们已经掌握 Android 应用程序中 Activity 的生命周期，了解了系统事件对 Activity 的生命周期的影响，以及如何保存与恢复 Activity 的状态。

3.6 习题与课外阅读

3.6.1 习题

(1) 分析当 Android 手机运行一个应用程序的 Activity 时，若收到短信，Activity 的生命周期变化。

(2) 编写一个程序演示 Activity 运行状态参数保存与恢复。

3.6.2 课外阅读

访问下列技术网站，拓展 Android 的 Activity 知识：

- http://www.oschina.net/question/54100_27841；
- http://www.oschina.net/android/65/android-activity。

第4章 用户界面的布局管理与视图

Android 应用程序的用户界面是与用户直接沟通的途径，Activity 是界面控件（View）的载体，每一个界面控件（如按钮、文本框等）不能直接摆放到 Activity 之上，而是通过布局管理器来管理如何放置的。因此在学习图形界面的设计之前，我们必须首先要掌握常见布局管理器的使用。本章主要介绍以下常用的五大布局管理器和两大视图：

- 线性布局（LinearLayout）；
- 相对布局（RelativeLayout）；
- 帧布局（FrameLayout）；
- 绝对布局（AbsoluteLayout）；
- 表格布局（TableLayout）；
- 列表视图（ListView）；
- 网格视图（GridView）。

本章学习目标：

掌握以上五大布局对控件的布局管理，以及两大视图显示方式效果及实现。

4.1 布局管理器的作用

由于 Android 系统支持的设备有平板电脑、智能手机、Android 手表等相关产品，这些设备的屏幕尺寸和分辨率可谓千差万别。为了保证一个 Android 程序能在不同的设备上以正确显示运行，一个设计良好的 Android 应用程序的用户界面应该能够自动适应不同设备，并做出明智的布局的变化，充分利用可用屏幕空间；必须考虑程序的逻辑业务与界面在不同设备上显示的兼容性，这方面的一个关键工作主要是依靠布局管理器来实现的。

4.2 View 和 ViewGroup 概述

在 Android 应用程序中，用户界面通过 View 和 ViewGroup 对象构建的。

- View 对象是 Android 平台上表示用户界面的基本单元；View 如何放置由 ViewGroup 管理；
- ViewGroup 中包含的一些 View 怎么样布局，View 的布局显示方式直接影响用户界面。

ViewGroup 类是布局（layout）和视图容器（View container）的基类，此类也定义了

ViewGroup.LayoutParams 类，它作为布局参数的基类，此类告诉父视图其中的子视图想如何显示。

ViewGroup 的子类有很多，如 AbsoluteLayout、AdapterView＜T extends Adapter＞、DrawerLayout、FragmentBreadCrumbs、FrameLayout、GridLayout、LinearLayout、PagerTitleStrip、RelativeLayout、SlidingDrawer、SlidingPaneLayout、SwipeRefreshLayout 和 ViewPager。

本章主要介绍常见的布局管理器：线性布局（LinearLayout）、相对布局（RelativeLayout）、表格布局（TableLayout）、标签布局（TabLayout）、绝对布局（AbsoluteLayout），以及两大视图：列表视图（ListView）和网格视图（GridView）。

综合使用这五种布局，可以在屏幕上将控件随心所欲地摆放，而且控件的大小和位置会随着屏幕大小的变化自动做出相应的调整。这五大布局、两大视图与 ViewGroup 和 View 类之间的关系如图 4-1 所示。

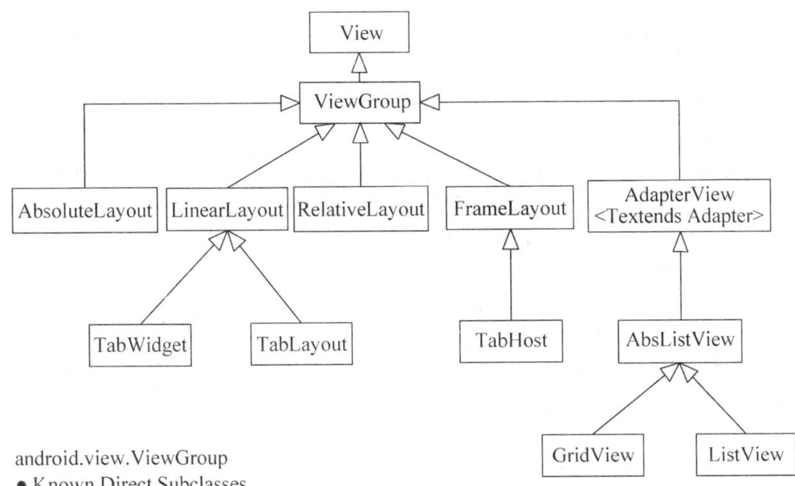

图 4-1　继承自 ViewGroup 的一些布局类

4.3　线性布局（LinearLayout）

线性布局是 Android 界面布局中比较常见的一种布局形式。线性布局将手机屏幕划分成一行或一列，布局管理器将可视化控件 View 按照线性顺序的方式摆放，超出屏幕的控件将不会显示出来。线性布局的形式可以分为两种。

- 第一种水平（横向）线性布局。设置线性布局为水平方向：

```
android:orientation = "horizontal"
```

• 第二种垂直(纵向)线性布局。设置线性布局为垂直方向：

```
android:orientation = "vertical"
```

如图 4-2 所示，可以清晰地看出来所有控件都是按照线性的排列方式显示出来的，这就是线性布局的特点。使用了线性布局的水平方向布局管理，放置了 5 个按钮，由于屏幕宽度空间所限，只完整地显示了 3 个按钮，第 4 个按钮 Button4 由于显示空间不足，按钮向纵向变形扩展，而第 5 个按钮 Button5，在屏幕之外，无法显示。图 4-2 的布局文件如下：

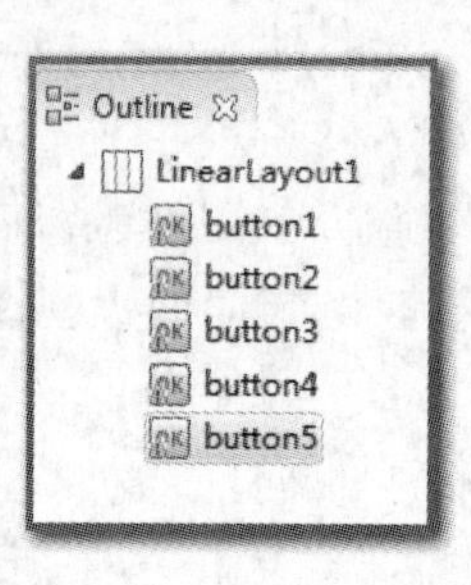

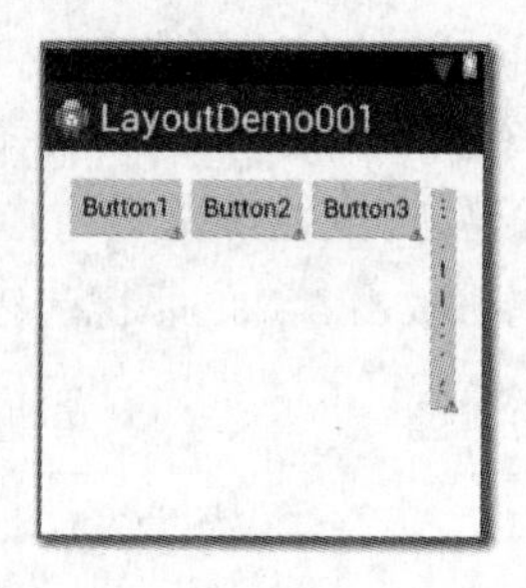

图 4-2　LinearLayout 布局

```
<LinearLayout xmlns:android = "http://schemas.android.com/apk/res/android"
    xmlns:tools = "http://schemas.android.com/tools"
    android:id = "@ + id/LinearLayout1"
    android:layout_width = "match_parent"
    android:layout_height = "match_parent"
    android:paddingBottom = "@dimen/activity_vertical_margin"
    android:paddingLeft = "@dimen/activity_horizontal_margin"
    android:paddingRight = "@dimen/activity_horizontal_margin"
    android:paddingTop = "@dimen/activity_vertical_margin"
    tools:context = ".MainActivity" >

    <Button
        android:id = "@ + id/button1"
        android:layout_width = "wrap_content"
        android:layout_height = "wrap_content"
        android:text = "Button1" />

    <Button
        android:id = "@ + id/button2"
        android:layout_width = "wrap_content"
        android:layout_height = "wrap_content"
        android:text = "Button2" />

    <Button
        android:id = "@ + id/button3"
        android:layout_width = "wrap_content"
        android:layout_height = "wrap_content"
        android:text = "Button3" />
```

```
    <Button
        android:id = "@ + id/button4"
        android:layout_width = "wrap_content"
        android:layout_height = "wrap_content"
        android:text = "Button4" />

    <Button
        android:id = "@ + id/button5"
        android:layout_width = "wrap_content"
        android:layout_height = "wrap_content"
        android:text = "Button5" />
</LinearLayout>
```

4.4 相对布局(RelativeLayout)

相对布局是 Android 布局中最为强人的一种布局结构,可视化控件的坐标取值范围都是相对的。Android 手机屏幕的分辨率可谓是千差万别,为了使得应用程序的界面能自适应屏幕的分辨率,所以在开发中建议大家都去使用相对布局,它的坐标取值范围都是相对的,所以使用它来实现自适应屏幕是正确的。图 4-3 和图 4-4 为采用相对布局的 5 个按钮。

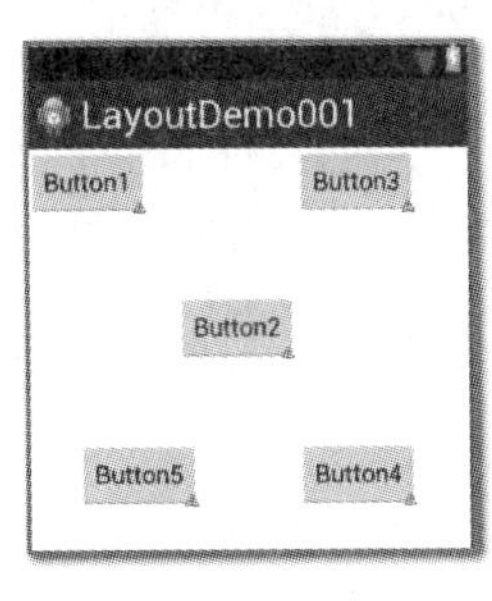

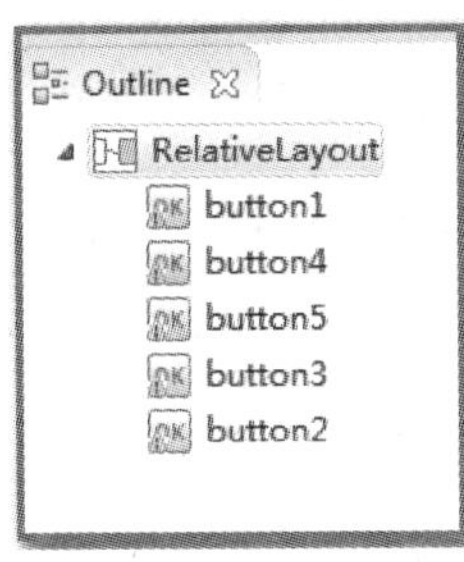

图 4-3 RelativeLayout 布局

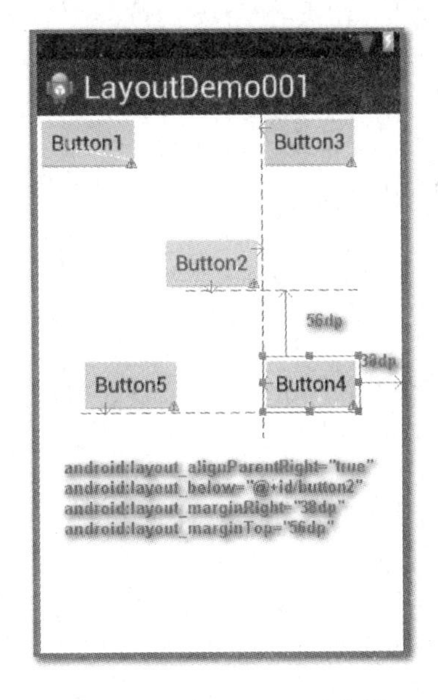

图 4-4 RelativeLayout 布局

布局管理文件如下:

```
<?xml version = "1.0" encoding = "utf - 8"?>
<RelativeLayout xmlns:android = "http://schemas.android.com/apk/res/android"
    android:layout_width = "match_parent"
    android:layout_height = "match_parent" >
```

```
    <Button
        android:id="@+id/button1"
        android:layout_width="wrap_content"
        android:layout_height="wrap_content"
        android:layout_alignParentLeft="true"
        android:layout_alignParentTop="true"
        android:text="Button1" />

    <Button
        android:id="@+id/button4"
        android:layout_width="wrap_content"
        android:layout_height="wrap_content"
        android:layout_alignParentRight="true"
        android:layout_below="@+id/button2"
        android:layout_marginRight="38dp"
        android:layout_marginTop="56dp"
        android:text="Button4" />

    <Button
        android:id="@+id/button5"
        android:layout_width="wrap_content"
        android:layout_height="wrap_content"
        android:layout_alignBottom="@+id/button4"
        android:layout_alignParentLeft="true"
        android:layout_marginLeft="36dp"
        android:text="Button5" />

    <Button
        android:id="@+id/button3"
        android:layout_width="wrap_content"
        android:layout_height="wrap_content"
        android:layout_above="@+id/button2"
        android:layout_alignLeft="@+id/button4"
        android:text="Button3" />

    <Button
        android:id="@+id/button2"
        android:layout_width="wrap_content"
        android:layout_height="wrap_content"
        android:layout_below="@+id/button1"
        android:layout_marginTop="55dp"
        android:layout_toLeftOf="@+id/button4"
        android:text="Button2" />

</RelativeLayout>
```

以 Button4 为例说明：

```
<Button
        android:id="@+id/button4"
```

```
        android:layout_width = "wrap_content"
        android:layout_height = "wrap_content"
        android:layout_alignParentRight = "true"
        android:layout_below = "@ + id/button2"
        android:layout_marginRight = "38dp"
        android:layout_marginTop = "56dp"
        android:text = "Button4" />
```

设置距父(button2)元素左、上对齐：

```
android:layout_alignParentLeft = "true" android:layout_alignParentTop = "true"
```

设置该控件在 id 为 button2 控件的下方：

```
android:layout_below = "@ + id/button2"
```

设置偏移的距离值

```
android:layout_marginRight = "38dp"
android:layout_marginTop = "56dp"
```

4.5 帧布局(FrameLayout)

该布局直接在屏幕上开辟出了一块空白区域，当我们往里面添加组件时，所有的组件都会放置于这块区域的左上角；帧布局的大小由子控件中最大的子控件决定，如果所有组件都一样大，同一时刻就只能看到最上面的那个组件了！帧布局在游戏开发方面用得比较多，例如可以使用图片做游戏场景的背景图。

```
<?xml version = "1.0" encoding = "utf - 8"?>
<FrameLayout xmlns:android = "http://schemas.android.com/apk/res/android"
    android:layout_width = "match_parent"
    android:layout_height = "match_parent" >

    <Button
        android:id = "@ + id/button"
        android:layout_width = "wrap_content"
        android:layout_height = "wrap_content"
        android:text = "Button - 001" />

    <Button
        android:id = "@ + id/button2"
        android:layout_width = "wrap_content"
        android:layout_height = "wrap_content"
        android:text = "B2" />

</FrameLayout>
```

如图 4-5 所示的界面中先绘制了 button，button 的 text 为 Button-001，然后又绘制了 button2，button2 的 text 为 B2，button2 覆盖了 button 的部分区域。

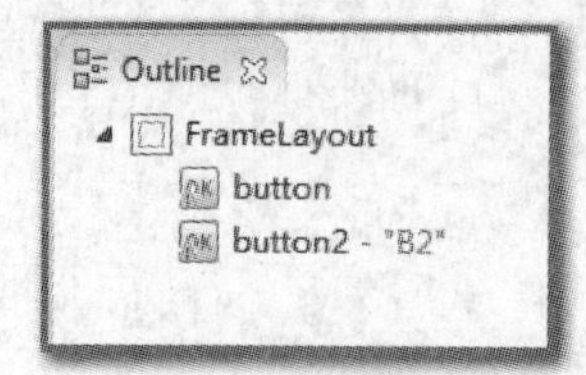

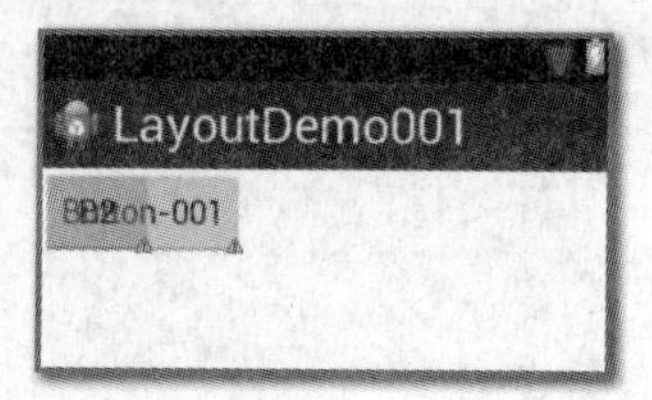

图 4-5 FrameLayout 布局

4.6 绝对布局(AbsoluteLayout)

绝对布局 AbsoluteLayout,又可以叫作坐标布局,使用绝对布局可以设置任意控件的在屏幕中的 X Y 绝对坐标值,如果两个控件所占据的空间有重叠,则和帧布局一样,后绘制的控件会覆盖住之前绘制的控件。这种布局简单直接,非常直观,但是由于手机屏幕尺寸差别比较大,使用绝对定位的适应性会比较差,很难保证在其他分辨率的手机上能正常显示了。除非针对特定的硬件开发(如工业控制显示面板),一般不提倡使用绝对布局来设计 UI。图 4-6 为绝对布局控制的三个按钮。

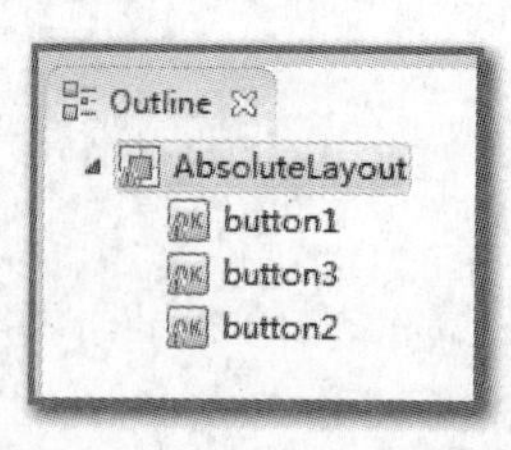

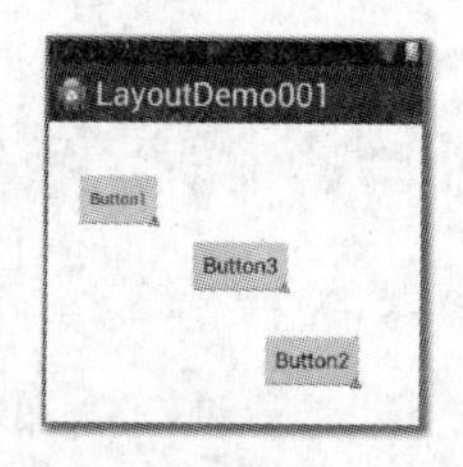

图 4-6 AbsoluteLayout 布局

```
<?xml version = "1.0" encoding = "utf - 8"?>
< AbsoluteLayout xmlns:android = "http://schemas.android.com/apk/res/android"
    android:layout_width = "match_parent"
    android:layout_height = "match_parent" >

    < Button
        android:id = "@ + id/button1"
        style = "?android:attr/buttonStyleSmall"
        android:layout_width = "wrap_content"
        android:layout_height = "wrap_content"
        android:layout_x = "24dp"
        android:layout_y = "40dp"
        android:text = "Button1" />

    < Button
        android:id = "@ + id/button3"
        android:layout_width = "wrap_content"
        android:layout_height = "wrap_content"
        android:layout_x = "122dp"
        android:layout_y = "95dp"
```

```
        android:text = "Button3" />

    <Button
        android:id = "@ + id/button2"
        android:layout_width = "wrap_content"
        android:layout_height = "wrap_content"
        android:layout_x = "184dp"
        android:layout_y = "173dp"
        android:text = "Button2" />

</AbsoluteLayout>
```

4.7 表格布局(TableLayout)

表格布局可以定义一系列的 TableRow 对象,用于行的显示。表格布局并不是表格,因此表格布局行、列和单元格不显示表格线。每个行可以包含 0 个以上(包括 0)的单元格,类似表格的列;每个单元格可以设置一个 View 对象。表格的单元格可以为空。

列的宽度由该列所有行中最宽的一个单元格决定。表格布局可以通过 setColumnShrinkable()方法或者 setColumnStretchable()方法来标记某些列可以收缩或可以拉伸。如果标记为可以收缩,列宽可以收缩以使表格适合容器的大小。如果标记为可以拉伸,那么列宽可以拉伸以占用多余的空间。表格的总宽度由其父容器决定。

列可以同时具有可拉伸和可收缩标记。在列可以调整其宽度以占用可用空间,但不能超过限度时是很有用的。可以通过调用 setColumnCollapsed()方法来隐藏列。表格布局的子对象宽度永远是 MATCH_PARENT,可以定义子对象的高度 layout_height 属性;其默认值是 WRAP_CONTENT。如果子对象是 TableRow,其高度永远是 WRAP_CONTENT。在表格布局中可以设置 TableRow 可以设置表格中每一行显示的内容以及位置,可以设置显示的缩进、对齐的方式(图 4-7 为表格布局控制的 11 个按钮)。

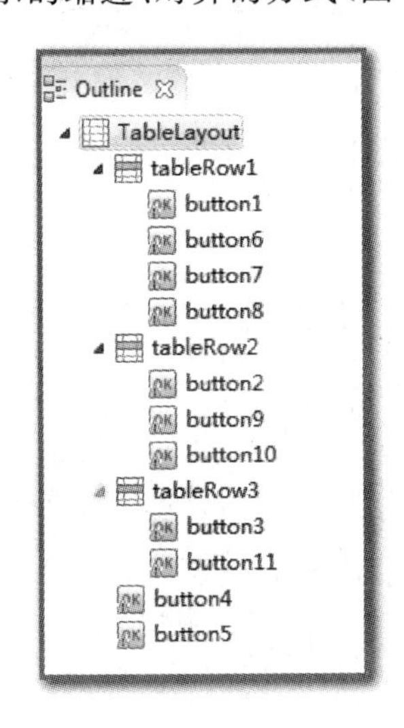

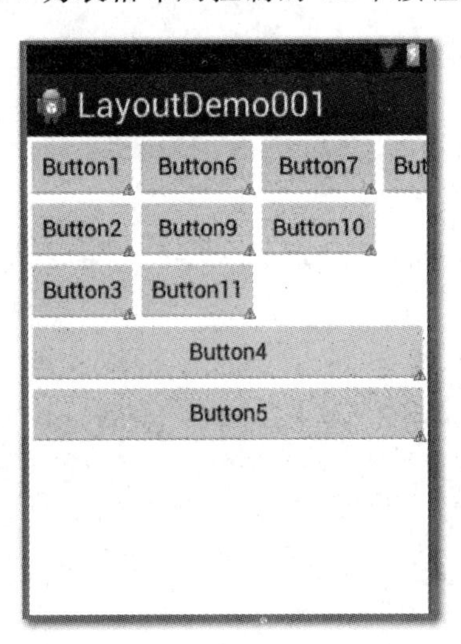

图 4-7 TableLayout 布局

```
<?xml version = "1.0" encoding = "utf - 8"?>
<TableLayout xmlns:android = "http://schemas.android.com/apk/res/android"
    android:layout_width = "match_parent"
    android:layout_height = "match_parent" >

    <TableRow
        android:id = "@ + id/tableRow1"
        android:layout_width = "wrap_content"
        android:layout_height = "wrap_content" >

        <Button
            android:id = "@ + id/button1"
            android:layout_width = "wrap_content"
            android:layout_height = "wrap_content"
            android:text = "Button1" />

        <Button
            android:id = "@ + id/button6"
            android:layout_width = "wrap_content"
            android:layout_height = "wrap_content"
            android:text = "Button6" />

        <Button
            android:id = "@ + id/button7"
            android:layout_width = "wrap_content"
            android:layout_height = "wrap_content"
            android:text = "Button7" />

        <Button
            android:id = "@ + id/button8"
            android:layout_width = "wrap_content"
            android:layout_height = "wrap_content"
            android:text = "Button8" />

    </TableRow>

    <TableRow
        android:id = "@ + id/tableRow2"
        android:layout_width = "wrap_content"
        android:layout_height = "wrap_content" >

        <Button
            android:id = "@ + id/button2"
            android:layout_width = "wrap_content"
            android:layout_height = "wrap_content"
            android:text = "Button2" />

        <Button
            android:id = "@ + id/button9"
```

```
            android:layout_width = "wrap_content"
            android:layout_height = "wrap_content"
            android:text = "Button9" />

        <Button
            android:id = "@ + id/button10"
            android:layout_width = "wrap_content"
            android:layout_height = "wrap_content"
            android:text = "Button10" />

    </TableRow>

    <TableRow
        android:id = "@ + id/tableRow3"
        android:layout_width = "wrap_content"
        android:layout_height = "wrap_content" >

        <Button
            android:id = "@ + id/button3"
            android:layout_width = "wrap_content"
            android:layout_height = "wrap_content"
            android:text = "Button3" />

        <Button
            android:id = "@ + id/button11"
            android:layout_width = "wrap_content"
            android:layout_height = "wrap_content"
            android:text = "Button11" />

    </TableRow>

    <Button
        android:id = "@ + id/button4"
        android:layout_width = "wrap_content"
        android:layout_height = "wrap_content"
        android:text = "Button4" />

    <Button
        android:id = "@ + id/button5"
        android:layout_width = "wrap_content"
        android:layout_height = "wrap_content"
        android:text = "Button5" />

</TableLayout>
```

Android 的五大布局各有自己的特点，其中相对布局是最强大的，其次它基本可以实现其他四大布局的效果；有时候纯使用一种布局不能满足复杂 UI 设计要求，通常会综合嵌套使用各种布局线。

4.8 列表视图(ListView)

列表视图的布局方式：是一个 ViewGroup 以列表显示它的子视图(view)元素，列表是可滚动的列表。列表元素通过 ListAdapter 自动插入到列表。布局文件中定义了 ListView，Adapter 用来将数据填充到 ListView，要填 ListView 的数据，可以是字符串、图片、控件等。其关系见图 4-8。

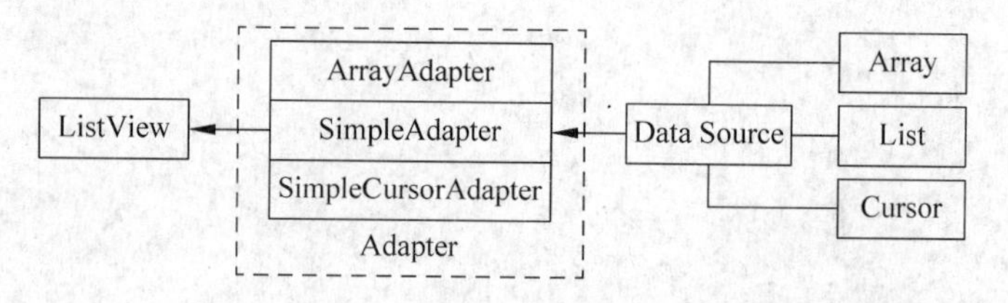

图 4-8 ListView 类与数据集之间关系

ListAdapter：扩展自 Adapter，它是 ListView 和数据源之间的桥梁。ListView 可以显示任何包装在 ListAdapter 中的数据。

根据数据源的不同 Adapter 可以分为三类：

(1) String[]：ArrayAdapter；

(2) List<Map<String,？>>：BaseAdapter；

(3) 数据库 Cursor：SimpleCursorAdapter。

使用 ArrayAdapter(数组适配器)顾名思义，需要把数据放入一个数组以便显示，上面的例子就是这样的；BaseAdapter 能定义各种各样的布局出来，可以放上 ImageView(图片)，还可以放上 Button(按钮)和 CheckBox(复选框)等；SimpleCursorAdapter 与数据库有关。

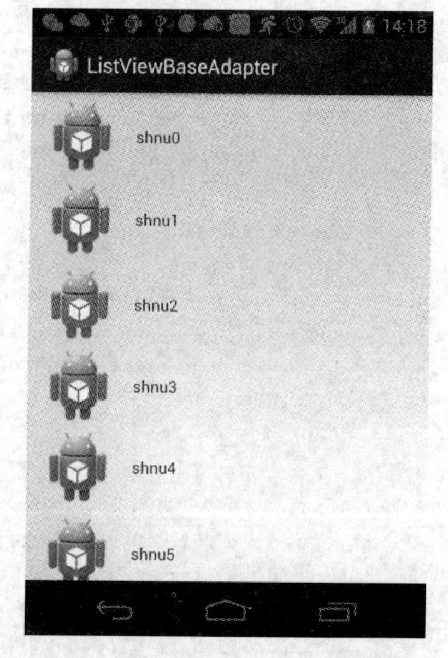

图 4-9 ListView 显示效果

图 4-9 为 ListView 的一个实例。

```
<?xml version = "1.0" encoding = "utf - 8"?>
< RelativeLayout xmlns:android = "http://schemas.android.com/apk/res/android"
    android:layout_width = "fill_parent"
    android:layout_height = "fill_parent"
    >
    < ListView
        android:id = "@ + id/listview"
        android:layout_width = "wrap_content"
        android:layout_height = "fill_parent"
        android:cacheColorHint = "#00000000"
        android:divider = "@null"
```

```
        android:drawSelectorOnTop = "false"
        android:scrollbars = "none"
        />
</RelativeLayout>
MainActivity.Java
package com.shnu;

import java.util.ArrayList;
import java.util.HashMap;
import java.util.List;
import java.util.Map;



import android.os.Bundle;
import android.view.View;
import android.widget.AdapterView;
import android.widget.ListView;
import android.widget.AdapterView.OnItemClickListener;
import android.app.Activity;

public class MainActivity extends Activity {
    private ListView listView;

    private List<Map<String, Object>> mData;

    private MyAdapter adapter;

    @Override
    protected void onCreate(Bundle savedInstanceState) {
        super.onCreate(savedInstanceState);
        setContentView(R.layout.activity_main);

        listView = (ListView) findViewById(R.id.listview);

       mData = new ArrayList<Map<String, Object>>();
        for (int i = 0; i < 10; i++) {
                Map<String, Object> map = new HashMap<String, Object>();
                map.put("School", "shnu" + i);
                mData.add(map);
        }
       adapter = new MyAdapter(this, mData);

       listView.setAdapter(adapter);

       listView.setOnItemClickListener(itemClick);
    }
```

```
    private OnItemClickListener itemClick = new OnItemClickListener()
    {
        @Override
        public void onItemClick(AdapterView<?> arg0, View arg1, int arg2,
                    long arg3) {
        }
    };

}

MyAdapter.java
package com.shnu;

import java.util.List;
import java.util.Map;

import android.content.Context;
import android.view.LayoutInflater;
import android.view.View;
import android.view.ViewGroup;
import android.widget.BaseAdapter;
import android.widget.ImageView;
import android.widget.TextView;

public class MyAdapter extends BaseAdapter {
    private LayoutInflater mInflater;

    private List<Map<String, Object>> list;

    public MyAdapter(Context context, List<Map<String, Object>> list) {
        this.list = list;
        mInflater = LayoutInflater.from(context);
    }

    private class Holder {
        ImageView img;
        TextView title;
    }

    @Override
    public int getCount() {
        return null == list ? 0 : list.size();
    }

    @Override
    public Object getItem(int position) {
        return list.get(position);
```

```
    }

    @Override
    public long getItemId(int position) {
        return position;
    }

    @Override
    public View getView(int position, View convertView, ViewGroup parent) {
        Holder holder = null;
        if (null == convertView) {
            holder = new Holder();
            convertView = mInflater.inflate(R.layout.tab_all_classify_item,
                        null);
            holder.img = (ImageView) convertView.findViewById(R.id.img);
            holder.title = (TextView) convertView.findViewById(R.id.title);
            convertView.setTag(holder);
        } else {
            holder = (Holder) convertView.getTag();
        }
        if (null != list && !list.isEmpty()) {
            if (null == list.get(position) && list.get(position).size() == 0) {
                    return convertView;
            }
            Map<String, Object> map = list.get(position);
            holder.img.setImageResource(R.drawable.ic_launcher);

            String appName = String.valueOf(map.get("School"));
            holder.title.setText(appName);
        } else {
            holder.title.setText("");
        }
        ;
        return convertView;
    }
}
```

4.9 网格视图(GridView)

网格视图的布局方式：是一个 ViewGroup 以网格显示它的子视图(view)元素，即二维的、滚动的网格。网格元素通过 ListAdapter 自动插入到网格。ListAdapter 跟上面的列表布局是一样的，此处不再赘述。

下面也通过一个例子来创建一个显示图片缩略图的网格。当一个元素被选择时，显示该元素在列表中的位置的消息(见图 4-10)。

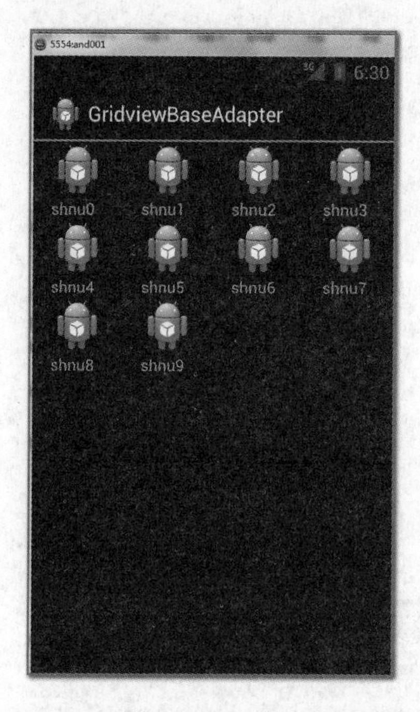

图 4-10　GridView 显示效果类与数据集之间的关系

```
<?xml version = "1.0" encoding = "utf - 8"?>
<RelativeLayout xmlns:android = "http://schemas.android.com/apk/res/android"
    android:layout_width = "fill_parent"
    android:layout_height = "fill_parent"
    >
    <GridView
        android:id = "@ + id/gridView"
        android:layout_width = "wrap_content"
        android:layout_height = "wrap_content"
        android:numColumns = "4"
        android:cacheColorHint = "#00000000"
        android:divider = "@null"
        android:drawSelectorOnTop = "false"
        android:scrollbars = "none"
        />
</RelativeLayout>
MainActivity.java
package com.shnu;

import java.util.ArrayList;
import java.util.HashMap;
import java.util.List;
import java.util.Map;
import android.os.Bundle;
import android.view.View;
import android.widget.AdapterView;
```

```java
import android.widget.GridView;
import android.widget.AdapterView.OnItemClickListener;
import android.app.Activity;

public class MainActivity extends Activity {
    private GridView gridView;

    private List<Map<String, Object>> mData;

    private MyAdapter adapter;

    @Override
    protected void onCreate(Bundle savedInstanceState) {
        super.onCreate(savedInstanceState);
        setContentView(R.layout.activity_main);

        gridView = (GridView) findViewById(R.id.gridView);

       mData = new ArrayList<Map<String, Object>>();
        for (int i = 0; i<10; i++) {
                Map<String, Object> map = new HashMap<String, Object>();
                map.put("School", "shnu" + i);
                mData.add(map);
        }
       adapter = new MyAdapter(this, mData);

       gridView.setAdapter(adapter);

       gridView.setOnItemClickListener(itemClick);
    }
    private OnItemClickListener itemClick = new OnItemClickListener()
    {
        @Override
        public void onItemClick(AdapterView<?> arg0, View arg1, int arg2,
                       long arg3) {
        }
    };

}

MyAdapter.java
package com.shnu;

import java.util.List;
import java.util.Map;

import android.content.Context;
import android.view.LayoutInflater;
import android.view.View;
```

```
import android.view.ViewGroup;
import android.widget.BaseAdapter;
import android.widget.ImageView;
import android.widget.TextView;
public class MyAdapter extends BaseAdapter
{
    private LayoutInflater mInflater;

    private List<Map<String, Object>> list;

    public MyAdapter(Context context,List<Map<String, Object>> list)
    {
    this.list = list;
        mInflater = LayoutInflater.from(context);
    }

    private class Holder
    {
        ImageView img;
       TextView title;
     }

    @Override
    public int getCount()
    {
        return null == list?0:list.size();
    }

    @Override
    public Object getItem(int position)
    {
         return list.get(position);
    }

    @Override
    public long getItemId(int position)
    {
         return position;
    }

    @Override
    public View getView(int position, View convertView, ViewGroup parent)
    {
        Holder holder = null;
        if (null == convertView)
        {
            holder = new Holder();
            convertView = mInflater.inflate(R.layout.tab_all_classify_item, null);
            holder.img = (ImageView) convertView
```

```
            .findViewById(R.id.img);
        holder.title = (TextView) convertView
                .findViewById(R.id.title);
        convertView.setTag(holder);
    }
    else
    {
        holder = (Holder) convertView.getTag();
    }
    if (null != list && !list.isEmpty())
      {
      if(null == list.get(position)
                        && list.get(position).size() == 0)
      {
            return convertView;
      }
      Map<String, Object> map = list.get(position);
            holder.img.setImageResource(R.drawable.ic_launcher);

            String appName = String.valueOf(map.get("School"));
            holder.title.setText(appName);
      }
      else
      {
            holder.title.setText("");
      };
    return convertView;
  }
}
```

4.10 本章小结

通过本章的学习，我们已经掌握了 Android 应用程序中五大布局、两大视图。

4.11 习题与课外阅读

4.11.1 习题

(1) 试分析图 4-11 中 QQ 聊天信息界面中的布局管理器与控件的位置。

(2) 试分析中国建设银行手机银行界面中的布局管理器与控件的位置(见图 4-12)。

4.11.2 课外阅读

(1) 访问下列技术网站，了解一下 Android SDK API 文档中 ViewGroup 的子类：DrawerLayout、FragmentBreadCrumbs、GridLayout、PagerTitleStrip、SlidingDrawer、SlidingPaneLayout、SwipeRefreshLayout、ViewPager 以及相关布局管理器的显示效果：

http://developer.android.com/

图 4-11　习题 1 要实现的界面

图 4-12　习题 2 要实现的界面

(2) 阅读文章"Android 开发中自定义 View 设定到 FrameLayout 布局中实现多组件显示"

http://www.oschina.net/code/snippet_4873_6234

第 5 章　Android 常见的 UI 控件

Android 应用程序的用户体验性非常重要，UI 设计是关键，因此必须熟练地掌握 Android 应用常见的 UI 控件的属性及事件处理机制。

本章学习目标：

- 掌握 Android 常见 UI 控件的特征、属性；
- 掌握 Android 常见 UI 控件的事件处理机制；
- 掌握如何用 Android 常见的 UI 可视化数据；
- 学会使用基本的 UI 控件编写程序。

5.1　Android 常见 UI 控件介绍

Android 系统中提供了丰富的 UI 控件，见图 5-1。

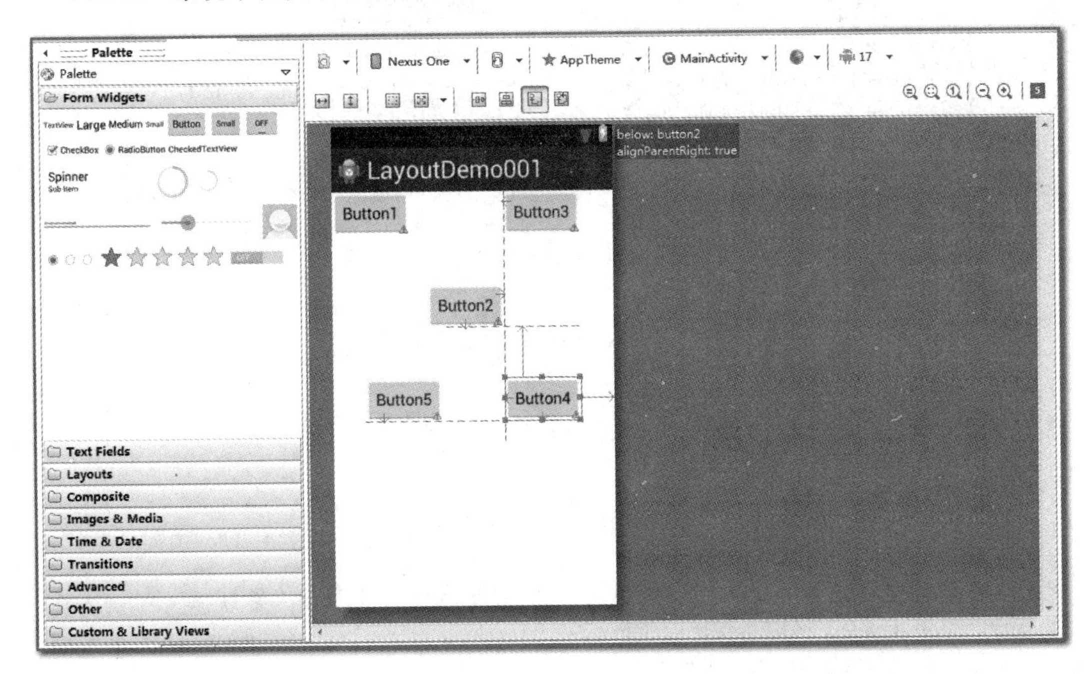

图 5-1　Android 常见 UI 一览图

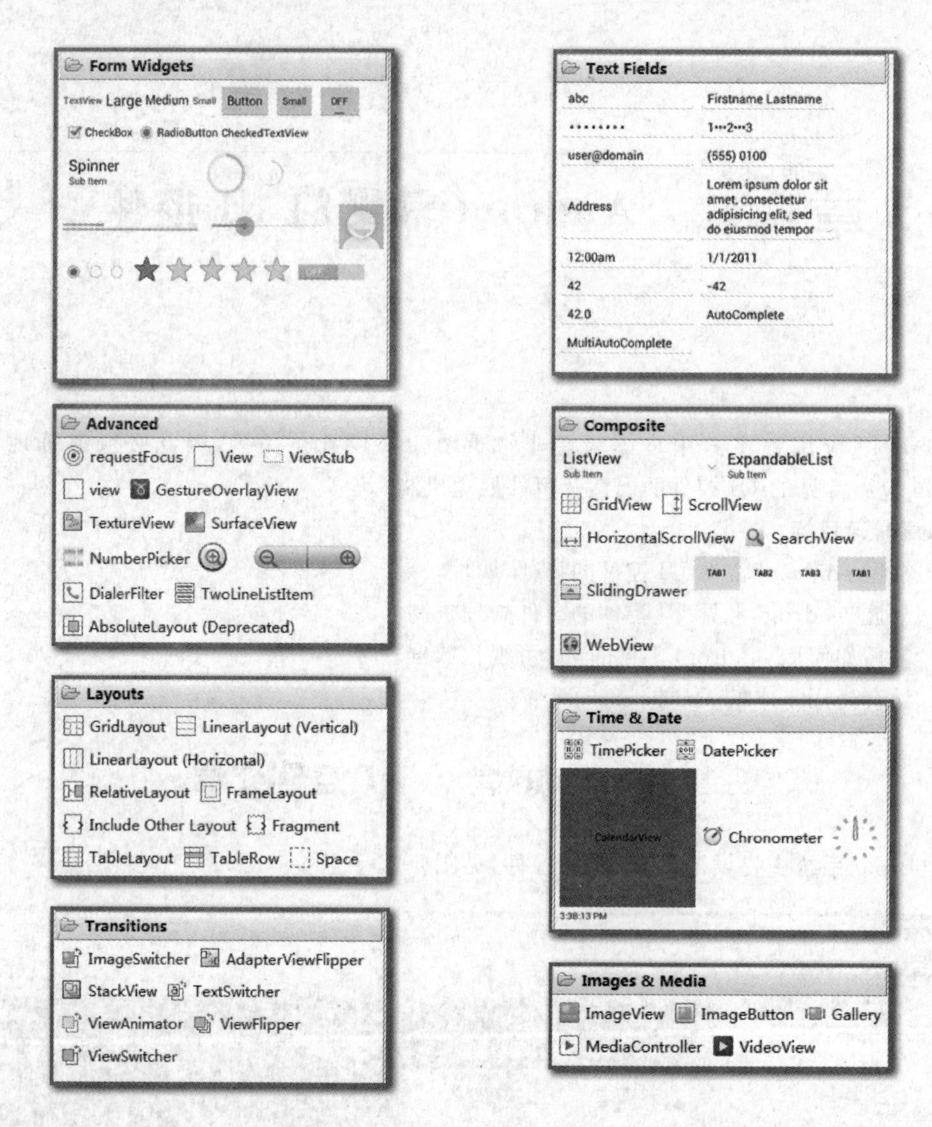

图 5-1 （续）

5.2 UI 控件的学习策略

对于 Android 提供的 UI 控件(View)的学习，可以采用以下学习策略：

① 任何控件都有 id，可以设置其状态数据；

② 要了解控件的外观和特征主要用途；

③ 要掌握如何利用控件可视化数据；

④ 要掌握如何捕获用户与控件的交互事件，并学会如何处理相关事件。

以下介绍 Button、ImageButton、Toast、TextView、EditText、CheckBox、RadioGroup Spinner、RatingBar 的使用。

5.3 Button 按钮

5.3.1 Button 类的结构

Button 类的层次关系如下：

```
java.lang.Object
        └android.view.View
            └android.widget.TextView
        └android.widget.Button
```

直接子类：

```
CompoundButton
```

间接子类：

```
CheckBox, RadioButton, ToggleButton
```

5.3.2 Button 常用的方法

Button 常用方法如表 5-1 所示。

表 5-1 Button 常用的方法

主要方法	功能描述	返回值
Button	Button 类的构造方法	Null
onKeyDown	当用户按键时，该方法调用	Boolean
onKeyUp	当用户按键弹起后，该方法被调用	Boolean
onKeyLongPress	当用户保持按键时，该方法被调用	Boolean
onKeyMultiple	当用户多次调用时，该方法被调用	Boolean
invalidateDrawable	刷新 Drawable 对象	void
scheduleDrawable	定义动画方案的下一帧	void
unscheduleDrawable	取消 scheduleDrawable 定义的动画方案	void
onPreDraw	设置视图显示，例如在视图显示之前调整滚动轴的边界	Boolean
sendAccessibilityEvent	发送事件类型指定的 AccessibilityEvent。发送请求之前，需要检查 Accessibility 是否打开	void
sendAccessibilityEventUnchecked	发送事件类型指定的 AccessibilityEvent。发送请求之前，不需要检查 Accessibility 是否打开	void
setOnKeyListener	设置按键监听	void

5.3.3 Button 标签的属性

Button 标签的属性如表 5-2 所示。

表 5-2　ButtonXML 属性

属性名称	描　述
android：layout_height	设置控件高度。可选值：fill_parent，warp_content，px
android:layout_width	设置控件宽度，可选值：fill_parent，warp_content，px
android：text	设置控件名称，可以是任意字符
android：layout_gravity	设置控件在布局中的位置，可选项：top、left、bottom、right、center_vertical、fill_vertica、fill_horizonal、center、fill 等
android：layout_weight	设置控件在布局中的比重，可选值：任意的数字
android：textColor	设置文字的颜色
android：bufferType	设置取得的文本类别，normal、spannable、editable
android：hint	设置文本为空是所显示的字符
android：textColorHighlight	设置文本被选中时，高亮显示的颜色
android：inputType	设置文本的类型，none、text、textWords 等
android:minWidth	设置文本区域的最小宽度

5.3.4　Button 的使用

Button 可以在 xml 中声明，也可以在代码中动态创建。

在 xml 中定义：

```
<Button
        android:id = "@ + id/button1"
        android:layout_width = "wrap_content"
        android:layout_height = "wrap_content"
        android:layout_alignParentLeft = "true"
        android:layout_alignParentTop = "true"
        android:text = "Button" />
```

效果见图 5-2 Button 运行界面。

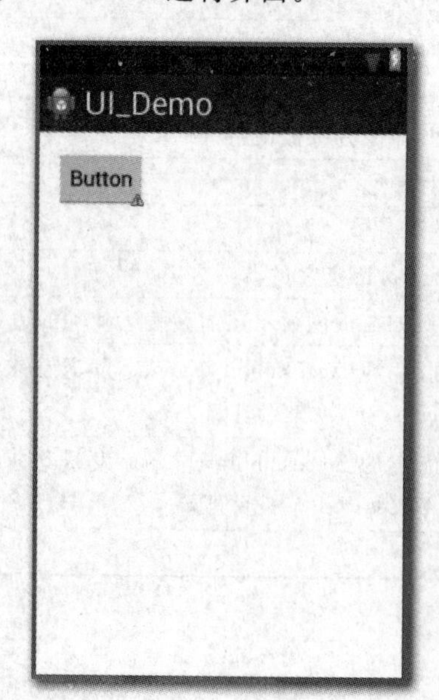

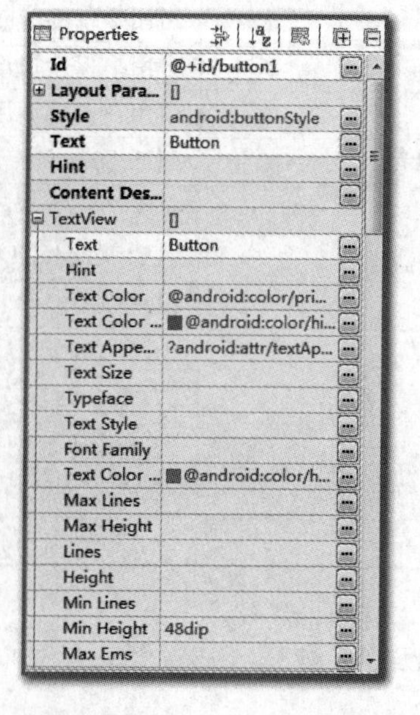

图 5-2　Button 效果图

创建好了 Button 后，就可以对其进行监听了，具体有两种方式。

- 一种是继承 OnClickListenner 接口：

```
public class ButtonActivity extends Activity implements OnClickListener{
    button1 = (Button)findViewById(R.id.button1);
    button1.setOnClickListener(this);
    public void onClick(View v) {
      switch(v.getId()){
          case R.id.button1:
            ⋮
          break;
    }
    }
    }
```

- 另一种方式是：

```
public class ButtonActivity extends Activity{
  button1 = (Button)findViewById(R.id.button1);
button1.setOnClickListener(new Button.OnClickListener(){

          public void onClick(View v) {
              // TODO Auto-generated method stub
              ⋮
          }

        });
```

5.4 ImageButton 按钮

5.4.1 ImageButton 类的结构

ImageButton 是带有图标的按钮，它的类层次关系如下：

```
java.lang.Object
   └android.view.View
     └android.widget.ImageView
       └android.widget.ImageButton
```

5.4.2 ImageButton 常用的方法

ImageButton 常用方法如表 5-3 所示。

5.4.3 ImageButton 标签的属性

ImageButton 标签属性如表 5-4 所示。

表 5-3 Android ImageButton 常用的方法

主要方法	功能描述	返回值
ImageButton	构造函数	null
setAdjustViewBounds	设置是否保持高宽比，需要与 maxWidth 和 maxHeight 结合起来一起使用	Boolean
getDrawable	获取 Drawable 对象，获取成功返回 Drawable，否则返回 null	Drawable
getScaleType	获取视图的填充方式	ScaleType
setScaleType	设置视图的填充方式，包括矩阵、拉伸等七种填充方式	void
setAlpha	设置图片的透明度	void
setMaxHeight	设置按钮的最大高度	void
setMaxWidth	设置按钮的最大宽度	void
setImageURI	设置图片的地址	void
setImageResource	设置图片资源库	void
setOnTouchListener	设置事件的监听	Boolean
setColorFilter	设置颜色过滤	void

表 5-4 Android ImageButton 标签的属性

属性名称	描述
android：adjustViewBounds	设置是否保持宽高比，True 或 False
android：cropToPadding	是否截取指定区域用空白代替。单独设置无效果，需要与 scrollY 一起使用，True 或者 False
android：maxHeight	设置图片按钮的最大高度
android：maxWidth	设置图片的最大宽度
android：scaleType	设置图片的填充方式
android：src	设置图片按钮的 drawable
android：tint	设置图片为渲染颜色
android：hint	设置文本为空是所显示的字符

5.4.4 ImageButton 的使用

ImageButton 可以采用两种方法创建。

其一，在 Xml 中声明，在 xml 中声明，在 xml 和代码中都可以实现，但相比较而言，在 xml 中实现更有利于代码的改动。

```
< ImageButton
        android:id = "@ + id/imageButton1"
        android:layout_width = "wrap_content"
        android:layout_height = "wrap_content"
        android:layout_alignLeft = "@ + id/button1"
        android:layout_below = "@ + id/button1"
        android:layout_marginTop = "28dp"
        android:src = "@drawable/ic_launcher" />
```

其二,在代码中创建:

```
imagebutton3 = new ImageButton(this);
imagebutton3.setId(100);
//设置自己的图片
imagebutton3.setBackgroundDrawable(getResources().getDrawable(R.drawable.p2));
```

接下来就可以对 imagebutton 进行监听,通过继承 OnClickListener 接口来实现。

```
        imagebutton1 = (ImageButton)findViewById(R.id.button1);
        //注册监听
        imagebutton1.setOnClickListener(this);
        //加入布局
        layout = new LinearLayout(this);
        layout.addView(imagebutton3,param);
        linnearlayout.addView(layout,param);
public void onClick(View v) {
        // TODO Auto - generated method stub
        switch(v.getId()){
        case R.id.button1:
                textveiw.setText("You click ImageButton1");
                break;
```

效果如图 5-3 所示。

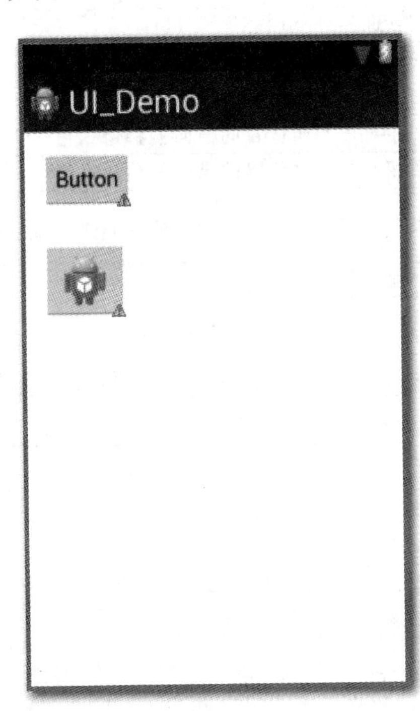

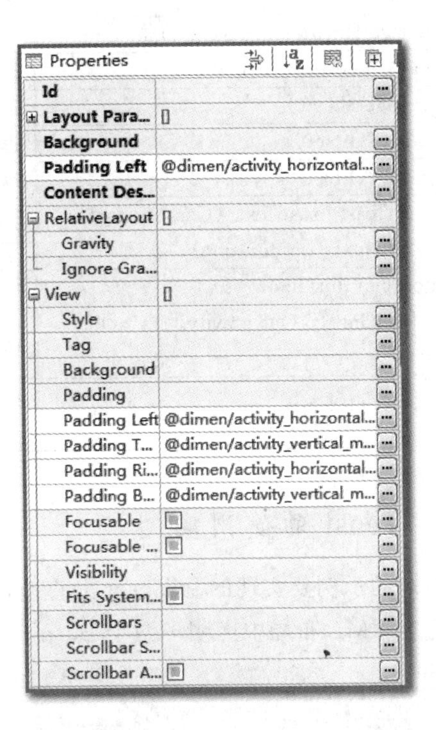

图 5-3 ImageButton 效果图及属性设置

5.5 Toast 提示

5.5.1 Toast 类的层次关系

Toast 是 Android 提供的"快显信息"类，它的用途很多，使用起来非常简单。

5.5.2 Toast 类常用的方法

Toast 中有两个关于 Toast 显示时间长短的常量，该时间长度可定制。参见 setDuration(int)（见表 5-5）。

- int LENGTH_LONG——持续显示视图或文本提示较长时间。
- int LENGTH_SHORT——持续显示视图或文本提示较短时间

表 5-5 Toast 常用的方法

主要方法	功能描述	返回值
public int cancel()	如果视图已经显示则将其关闭，还没有显示则不再显示。一般不需要调用该方法。正常情况下，视图会在超过存续期间后消失	void
public int getDuration()	返回存续期间	void
int getGravity()	取得提示信息在屏幕上显示的位置。 请参阅 API 文档	int
public float getHorizontalMargin ()	返回横向栏外空白	float
public float getVerticalMargin()	返回纵向栏外空白	float
public View getView()	返回 View 对象。请参阅 API 文档	View
public int getXOffset()	返回相对于参照位置的横向偏移像素量	int
public int getYOffset ()	返回相对于参照位置的纵向偏移像素量	int
public static Toast makeText (Context context, CharSequence text, int duration)	生成一个包含文本视图的标准 Toast 对象	Toast
setDuration (int duration)	设置存续期间	void
public void setGravity (int gravity, int xOffset, int yOffset)	设置提示信息在屏幕上的显示位置	void
public void setText (CharSequence s)	更新之前通过 makeText() 方法生成的 Toast 对象的文本内容	void

5.5.3 Toast 的使用实例

接下来的示例要实现的是 Toast 的直接显示以及 Toast 显示 View 的内容：

首先在 XML 布局中声明了两个 Button 按钮：

```
<Button android:id = "@ + id/button1"
    android:layout_width = "fill_parent"
    android:layout_height = "wrap_content"
    android:text = "Toast 显示 View"
```

```
/>
<Button android:id = "@ + id/button2"
    android:layout_width = "fill_parent"
    android:layout_height = "wrap_content"
    android:text = "Toast 直接输出"
/>
```

MainActivity.java 代码如下：

```
package com.example.ui_demo;
import android.os.Bundle;
import android.app.Activity;
import android.content.Context;
import android.view.LayoutInflater;
import android.view.Menu;
import android.view.View;
import android.view.View.OnClickListener;
import android.widget.Button;
import android.widget.Toast;

public class MainActivity extends Activity {

    @Override
    protected void onCreate(Bundle savedInstanceState) {
        super.onCreate(savedInstanceState);
        setContentView(R.layout.activity_main);
        Button button1 = (Button) findViewById(R.id.button1);
        button1.setOnClickListener(button1Listener);
        Button button2 = (Button) findViewById(R.id.button2);
        button2.setOnClickListener(button2Listener);

    }

    OnClickListener button1Listener = new OnClickListener() {

        public void onClick(View v) {
            showToast();
        }
    };
    OnClickListener button2Listener = new OnClickListener() {
        public void onClick(View v) {
            Toast.makeText(MainActivity.this, "直接输出", Toast.LENGTH_LONG)
                        .show();
        }
    };

    public void showToast() {
        LayoutInflater li = (LayoutInflater) getSystemService(Context.LAYOUT_INFLATER_SERVICE);
        View view = li.inflate(R.layout.toast, null);
```

```
            // 把布局文件 toast.xml 转换成一个 view
            Toast toast = new Toast(this);
            toast.setView(view);
            // 载入 view,即显示 toast.xml 的内容

            toast.setDuration(Toast.LENGTH_SHORT);
            // 设置显示时间,长时间 Toast.LENGTH_LONG,短时间为 Toast.LENGTH_SHORT,不可以自己编辑
            toast.show();
        }

    }
```

实现的效果图如图 5-4 所示。

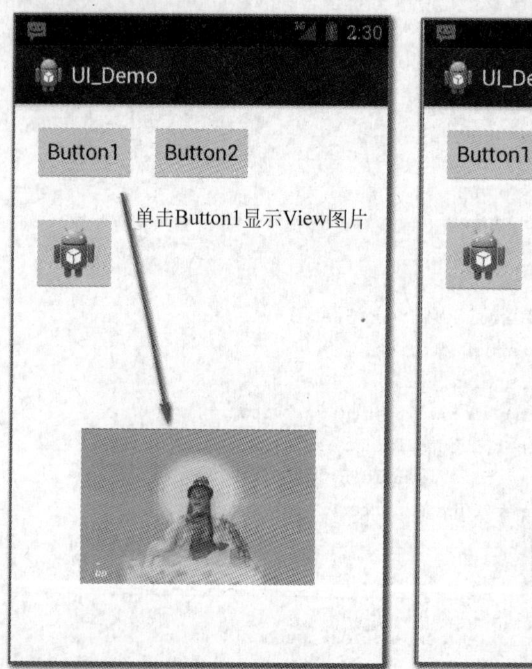

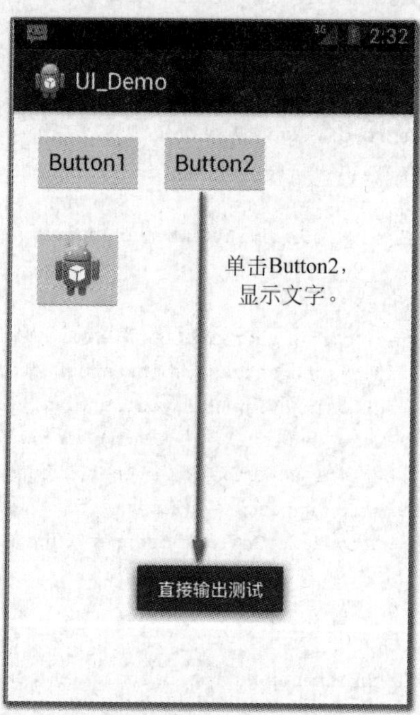

图 5-4　Toast 效果图

5.6　TextView 文本框

5.6.1　TextView 类的结构

TextView 是用于显示字符串的组件,对于用户来说就是屏幕中一块用于显示文本的区域。TextView 类的层次关系如下:

```
java.lang.Object
    └android.view.View
    └android.widget.TextView
```

直接子类：

```
Button, CheckedTextView, Chronometer, DigitalClock, EditText
```

间接子类：

```
AutoCompleteTextView, CheckBox, CompoundButton, ExtractEditText, MultiAutoCompleteTextView,
RadioButton, ToggleButton
```

5.6.2 TextView 类的方法

TextView 类的方法如表 5-6 所示。

表 5-6 TextView 类的方法

主要方法	功能描述	返回值
getDefaultMovementmethod	获取默认的箭头按键移动方式	Movementmethod
getText	获得 TextView 对象的文本	CharSquence
length	获得 TextView 中的文本长度	Int
getEditableText	取得文本的可编辑对象，通过 这个对象可对 TextView 的文本进行操作，如在光标之后插入字符	Void
getCompoundPaddingBottom	返回底部填充物	Int
setCompoundDrawables	设置图像显示的位置，在 设置该 Drawable 资源之前需要调用 setBounds(Rect)	Void
trinsicBounds	设置 Drawable 图像的显示位置，但其边界不变	Void
setPadding	根据位置设置填充物	Void
getAutoLinkMask	返回自动连接的掩码	Void
setTextColor	设置文本显示的颜色	Void
setHighlightColor	设置文本选中时显示的颜色	Void
setShadowLayer	设置文本显示的阴影颜色	Void
setHintTextColor	设置提示文字的颜色	Void
setLinkTextColor	设置链接文字的颜色	Void
setGravity	设置当 TextView 超出了文本本身时横向以及垂直对齐	Void
getFreezesText	设置该视图是否包含整个文本，如果包含则返回真值，否则返回假值	Boolean
TextView	TextView 的构造方法	Null
getDefaultMovementmethod	获取默认的箭头按键移动方式	Movementmethod

5.6.3 TextView 标签的属性

TextView 标签的属性如表 5-7 所示。

表 5-7 TextView 类的属性

属性名称	描述
android:autoLink	设置是否当文本为 URL 链接/email/电话号码/map 时,文本显示为可点击的链接。可选值(none/web/email/phone/map/all)
android:autoText	如果设置,将自动执行输入值的拼写纠正。此处无效果,在显示输入法并输入的时候起作用
android:bufferType	指定 getText( )方式取得的文本类别。选项 editable 类似于 StringBuilder 可追加字符,也就是说 getText 后可调用 append 方法设置文本内容
android:capitalize	设置英文字母大写类型。此处无效果,需要弹出输入法才能看得到,参见 API 文档
android:cursorVisible	设定光标为显示/隐藏,默认为显示
android:digits	设置允许输入哪些字符。如"1234567890.+－*/%\n()"
android:drawableBottom	在 text 的下方输出一个 drawable。如果指定一个颜色的话会把 text 的背景设为该颜色,并且同时和 background 使用时覆盖后者
android:drawableLeft	在 text 的左边输出一个 drawable
android:drawablePadding	设置 text 与 drawable(图片)的间隔,与 drawableLeft、drawableRight、drawableTop、drawableBottom 一起使用,可设置为负数,单独使用没有效果
android:drawableRight	在 text 的右边输出一个 drawable
android:drawableTop	在 text 的正上方输出一个 drawable
android:editable	设置是否可编辑。这里无效果,参见 API 文档
android:editorExtras	设置文本的额外的输入数据
android:ellipsize	设置当文字过长时,该控件该如何显示。有如下值设置:start 省略号显示在开头;end 省略号显示在结尾;middle 省略号显示在中间;marquee 以跑马灯的方式显示动画横向移动
android:freezesText	设置保存文本的内容以及光标的位置
android:gravity	设置文本位置,如设置成 center,文本将居中显示
android:hint	Text 为空时显示的文字提示信息,可通过 textColorHint 设置提示信息的颜色
android:linksClickable	设置链接是否可单击,即使设置了 autoLink
android:marqueeRepeatLimit	在 ellipsize 指定 marquee 的情况下,设置重复滚动的次数,当设置为 marquee_forever 时表示无限次
android:ems	设置 TextView 的宽度为 N 个字符的宽度。这里测试为一个汉字字符宽度
android:maxEms	设置 TextView 的宽度为最长为 N 个字符的宽度。与 ems 同时使用时覆盖 ems 选项
android:minEms	设置 TextView 的宽度为最短为 N 个字符的宽度。与 ems 同时使用时覆盖 ems 选项
android:maxLength	限制显示的文本长度,超出部分不显示

续表

属性名称	描　述
android:lines	设置文本的行数,设置两行就显示两行,即使第二行没有数据
android:maxLines	设置文本的最大显示行数,与 width 或者 layout_width 结合使用,超出部分自动换行,超出行数将不显示
android:minLines	设置文本的最小行数,与 lines 类似
android:lineSpacingExtra	设置行间距
android:lineSpacingMultiplier	设置行间距的倍数。如 1.2
android:numeric	如果被设置,该 TextView 有一个数字输入法。
android:password	以小点“.”显示文本
android:phoneNumber	设置为电话号码的输入方式
android:scrollHorizontally	设置文本超出 TextView 的宽度的情况下,是否出现横拉条
android:selectAllOnFocus	如果文本是可选择的,让它获取焦点而不是将光标移动为文本的开始位置或者末尾位置
android:singleLine	设置单行显示。如果和 layout_width 一起使用,当文本不能全部显示时,后面用“…”来表示。如 android:text=“test_ singleLine” android:singleLine=“true” android:layout_width=“20dp”将只显示“t…”。如果不设置 singleLine 或者设置为 false,文本将自动换行
android:text	设置显示文本
android:textColor	设置文本颜色
android:textColorHighlight	被选中文字的底色,默认为蓝色
android:textColorHint	设置提示信息文字的颜色,默认为灰色。与 hint 一起使用
android:textColorLink	文字链接的颜色
android:textScaleX	设置文字之间间隔,默认为 1.0f
android:textSize	设置文字大小,推荐度量单位为 sp,如 15sp
android:textStyle	设置字形[bold(粗体) 0, italic(斜体) 1, bolditalic(又粗又斜) 2] 可以设置一个或多个,用“|”隔开
android:typeface	设置文本字体,必须是以下常量值之一: normal 0、sans 1、serif 2、monospace(等宽字体) 3
android:height	设置文本区域的高度,支持度量单位: px(像素)/dp/sp/in/mm(毫米)
android:maxHeight	设置文本区域的最大高度
android:minHeight	设置文本区域的最小高度
android:width	设置文本区域的宽度,支持度量单位: px(像素)/dp/sp/in/mm(毫米)
android:maxWidth	设置文本区域的最大宽度
android:minWidth	设置文本区域的最小宽度

5.6.4 TextView 的使用

既可以在 XML 布局文件中声明及设置 TextView,也可以在代码中生成 TextView 组件。其效果图见图 5-5。

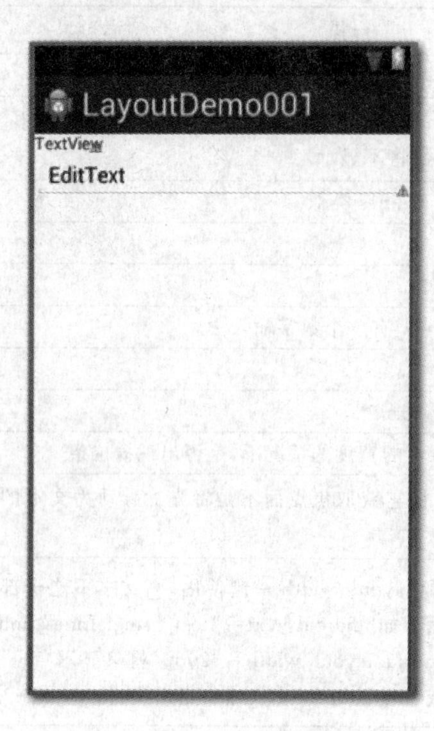

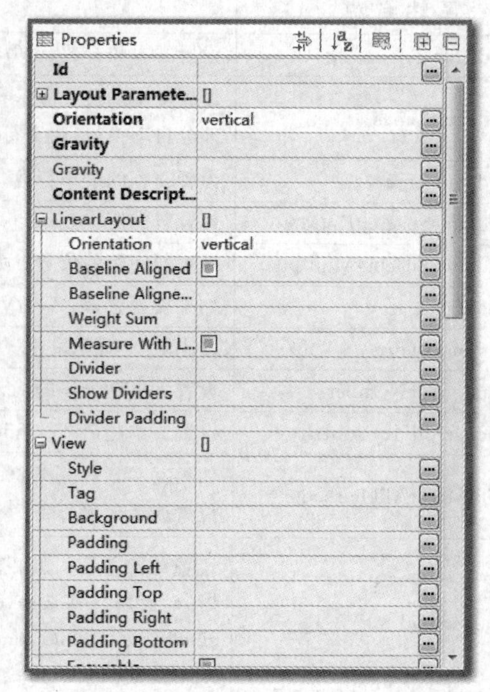

图 5-5　TextView 效果图

```
<?xml version = "1.0" encoding = "utf - 8"?>
< LinearLayout xmlns:android = "http://schemas.android.com/apk/res/android"
    android:layout_width = "match_parent"
    android:layout_height = "match_parent"
    android:orientation = "vertical" >

    < TextView
        android:id = "@ + id/textView1"
        android:layout_width = "wrap_content"
        android:layout_height = "wrap_content"
        android:text = "TextView" />

    < EditText
        android:id = "@ + id/editText1"
        android:layout_width = "match_parent"
        android:layout_height = "wrap_content"
        android:ems = "10"
        android:text = "EditText" />

</LinearLayout >
```

下面给出相关测试代码：

TextView. java：

```
  package com.shnu.view;
import android.app.Activity;
import android.os.Bundle;
import android.widget.TextView;
public class _TextView extends Activity {
    @Override
    protected void onCreate(Bundle savedInstanceState) {
        // TODO Auto-generated method stub
        super.onCreate(savedInstanceState);
        this.setContentView(R.layout.tt);
        TextView txt = (TextView) this.findViewById(R.id.textView1);
EditText etxt = (EditText) this.findViewById(R.id. editText1);

        // 设置文本显示控件的文本内容
        txt.setText("This TextView\n ");
etxt.setText("This EditText\n ");

    }
}
```

在代码中动态创建 TextView：

```
  linearLayout = (LinearLayout)findViewById(R.id.linearLayout01);
param = new LinearLayout.LayoutParams(LinearLayout.LayoutParams.FILL_PARENT,
                    LinearLayout.LayoutParams.WRAP_CONTENT);
    LinearLayout layout1 = new LinearLayout(this);
    layout1.setOrientation(LinearLayout.HORIZONTAL);
    TextView tv = new TextView(this);
    tv.setId(200);
    tv.setText("用代码动态创建 TextView");
    tv.setBackgroundColor(Color.GREEN);
    tv.setTextColor(Color.RED);
    tv.setTextSize(20);
    layout1.addView(tv, param);
    linearLayout.addView(layout1, param);
```

5.7 EditText 编辑框

5.7.1 EditText 类的结构

EditText 和 TextView 的功能基本类似，它们之间的主要区别在于 EditText 提供了可编辑的文本框。EditText 类关系如下：

```
java.lang.Object
    android.view.View
```

```
android.widget.TextView
    android.widget.EditText
```


直接子类：

```
AutoCompleteTextView, ExtractEditText
```

间接子类：

```
MultiAutoCompleteTextView
```

5.7.2 EditText 常用的方法

EditText 的常用方法如表 5-8 所示。

表 5-8 EditText 常用的方法

主要方法	功能描述	返回值
setImeOptions	设置软键盘的 Enter 键	void
getImeActionLable	设置 IME 动作标签	Charsequence
getDefaultEditable	获取是否默认可编辑	boolean
setEllipse	设置文件过长时控件的显示方式	void
setFreeezesText	设置保存文本内容及光标位置	void
getFreeezesText	获取保存文本内容及光标位置	boolean
setGravity	设置文本框在布局中的位置	void
getGravity	获取文本框在布局中的位置	int
setHint	设置文本框为空时，文本框默认显示的字符	void
getHint	获取文本框为空时，文本框默认显示的字符	Charsequence
setIncludeFontPadding	设置文本框是否包含底部和顶端的额外空白	void
setMarqueeRepeatLimit	在 ellipsize 指定 marquee 的情况下，设置重复滚动的次数，当设置为 marquee_forever 时表示无限次	void

5.7.3 EditText 标签的属性

EditText 标签的属性如表 5-9 所示。

表 5-9 EditText 常用的属性

属性名称	描述
android:autoLink	设置是否当文本为 URL 链接/email/电话号码/map 时，文本显示为可点击的链接。可选值(none/web/email/phone/map/all)。这里只有在同时设置 text 时才自动识别链接，后来输入的无法自动识别
android:autoText	自动拼写帮助
android:bufferType	指定 getText( )方式取得的文本类别。选项 editable 类似于 StringBuilder 可追加字符，也就是说 getText 后可调用 append 方法设置文本内容。spannable 则可在给定的字符区域使用样式

续表

属性名称	描　述
android:capitalize	设置英文字母大写类型。设置如下值：sentences 仅第一个字母大写；words 每一个单词首字母大小，用空格区分单词；characters 每一个英文字母都大写。在模拟器上用 PC 键盘直接输入可以出效果，但是用软键盘无效果
android:cursorVisible	设定光标为显示/隐藏，默认显示。如果设置 false，即使选中了也不显示光标栏
android:digits	设置允许输入哪些字符。如“1234567890. +－ * /%\n()”
android:drawableTop	在 text 的正上方输出一个 drawable
android:drawableBottom	在 text 的下方输出一个 drawable(如图片)。如果指定一个颜色的话会把 text 的背景设为该颜色，并且同时和 background 使用时覆盖后者
android:drawableLeft	在 text 的左边输出一个 drawable(如图片)
android:drawablePadding	设置 text 与 drawable(图片)的间隔，与 drawableLeft、drawableRight、drawableTop、drawableBottom 一起使用，可设置为负数，单独使用没有效果
android:drawableRight	在 text 的右边输出一个 drawable，如图片
android:editable	设置是否可编辑。仍然可以获取光标，但是无法输入
android:editorExtras	指定特定输入法的扩展，如“com. mydomain. im. SOME_FIELD”
android:ellipsize	设置当文字过长时，该控件该如何显示。有如下值设置：start——省略号显示在开头；end——省略号显示在结尾；middle——省略号显示在中间；marquee——以跑马灯的方式显示(动画横向移动)
android:freezesText	设置保存文本的内容以及光标的位置
android:gravity	设置文本位置，如设置成 center，文本将居中显示
android:hint	Text 为空时显示的文字提示信息，可通过 textColorHint 设置提示信息的颜色
android:imeOptions	设置软键盘的 Enter 键。有如下值可设置：normal、actionUnspecified、actionNone、actionGo、actionSearch、actionSend、actionNext、actionDone、flagNoExtractUi、flagNoAccessoryAction、flagNoEnterAction。可用“|”设置多个
android:imeActionId	设置 IME 动作 ID，在 onEditorAction 中捕获判断进行逻辑操作
android:imeActionLabel	设置 IME 动作标签。但是不能保证一定会使用，猜想在输入法扩展的时候应该有用
android:includeFontPadding	设置文本是否包含顶部和底部额外空白，默认为 True
android:inputMethod	为文本指定输入法，需要完全限定名(完整的包名)。例如：com. google. android. inputmethod. pinyin，但是这里报错找不到。关于自定义输入法参见这里。sentences 仅第一个字母大写；words 每一个单词首字母大小，用空格区分单词；characters 每一个英文字母都大写
android:inputType	设置文本的类型，用于帮助输入法显示合适的键盘类型
android:marqueeRepeatLimit	在 ellipsize 指定 marquee 的情况下，设置重复滚动的次数，当设置为 marquee_forever 时表示无限次
android:ems	设置 TextView 的宽度为 N 个字符的宽度。参见 TextView 中此属性的截图

续表

属性名称	描　述
android:maxEms	设置 TextView 的宽度为最长为 N 个字符的宽度。与 ems 同时使用时覆盖 ems 选项
android:minEms	设置 TextView 的宽度为最短为 N 个字符的宽度。与 ems 同时使用时覆盖 ems 选项
android:maxLength	限制输入字符数。如设置为 5,那么仅可以输入 5 个汉字/数字/英文字母
android:lines	设置文本的行数,设置两行就显示两行,即使第二行没有数据
android:maxLines	设置文本的最大显示行数,与 width 或者 layout_width 结合使用,超出部分自动换行,超出行数将不显示
android:minLines	设置文本的最小行数,与 lines 类似
android:linksClickable	设置链接是否单击连接,即使设置了 autoLink
android:lineSpacingExtra	设置行间距
android:lineSpacingMultiplier	设置行间距的倍数。如 1.2
android:numeric	如果被设置,该 TextView 有一个数字输入法。有如下值设置:integer 正整数、signed 带符号整数、decimal 带小数点浮点数
android:password	以小点"."显示文本
android:phoneNumber	设置为电话号码的输入方式
android:privateImeOptions	提供额外的输入法选项(字符串格式)。依据输入法而决定是否提供,如这里所见。自定义输入法继承 InputMethodService
android:scrollHorizontally	设置文本超出 TextView 的宽度的情况下,是否出现横拉条
android:selectAllOnFocus	如果文本是可选择的,让它获取焦点而不是将光标移动为文本的开始位置或者末尾位置。TextView 中设置后无效果
android:shadowColor	指定文本阴影的颜色,需要与 shadowRadius 一起使用
android:shadowDx	设置阴影横向坐标开始位置
android:shadowDy	设置阴影纵向坐标开始位置
android:shadowRadius	设置阴影的半径。设置为 0.1 就变成字体的颜色了,一般设置为 3.0 的效果比较好
android:singleLine	设置单行显示。如果和 layout_width 一起使用,当文本不能全部显示时,后面用"…"来表示。如 android:text="test_ singleLine" android:singleLine="true" android:layout_width="20dp"将只显示"t…"。如果不设置 singleLine 或者设置为 false,文本将自动换行
android:text	设置显示文本
android:textAppearance	设置文字外观
android:textColor	设置文本颜色
android:textColorHighlight	被选中文字的底色,默认为蓝色
android:textColorHint	设置提示信息文字的颜色,默认为灰色。与 hint 一起使用
android:textColorLink	文字链接的颜色
android:textScaleX	设置文字缩放,默认为 1.0f。参见 TextView 的截图
android:textSize	设置文字大小,推荐度量单位为 sp,如 15sp
android:textStyle	设置字形[bold(粗体) 0, italic(斜体) 1, bolditalic(又粗又斜) 2] 可以设置一个或多个,用"|"隔开

续表

属性名称	描　述
android:typeface	设置文本字体,必须是以下常量值之一: normal 0, sans 1, serif 2, monospace(等宽字体) 3
android:height	设置文本区域的高度,支持度量单位: px(像素)/dp/sp/in/mm(毫米)
android:maxHeight	设置文本区域的最大高度
android:minHeight	设置文本区域的最小高度
android:width	设置文本区域的宽度,支持度量单位: px(像素)/dp/sp/in/mm(毫米)
android:maxWidth	设置文本区域的最大宽度
android:minWidth	设置文本区域的最小宽度

5.7.4 EditText 的使用

EditText 和 TextView 一样,既可以在 Xml 中声明实现,也可以在代码中动态生成,关于在代码动态生成和 TextView 的类似,此处不再赘述。下面以一个实例来说明 EditText 的简单使用。样例效果见图 5-6。

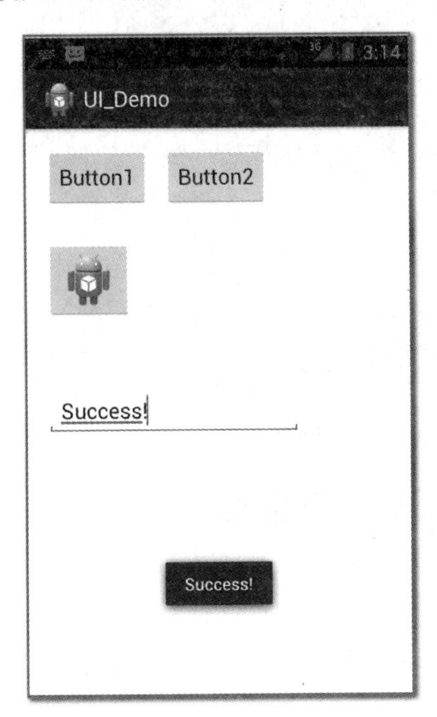

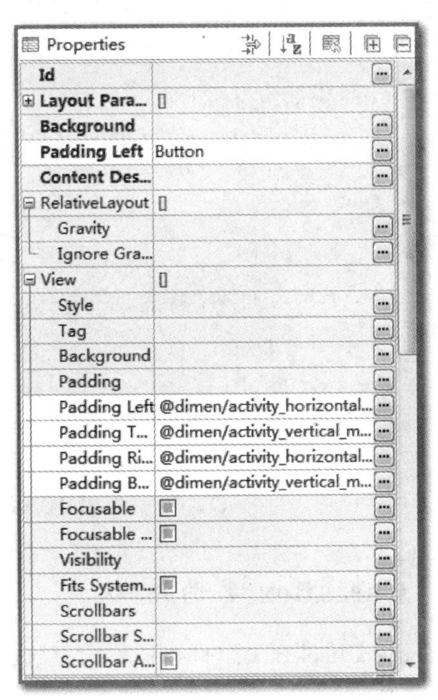

图 5-6　EditText 样例图

在 XML 布局中:

```
<EditText
        android:id = "@ + id/editText1"
        android:layout_width = "wrap_content"
        android:layout_height = "wrap_content"
```

```
        android:layout_alignLeft = "@ + id/imageButton1"
        android:layout_centerVertical = "true"
        android:ems = "10"
        android:hint = "hello" >

        < requestFocus />
    </EditText >
```

在 Activity 中让 EditText 显示在屏幕上并实现监听：

```
// 获得 EditTextView 对象
     edittext_num = (EditText) findViewById(R.id.editText1);
        edittext_num.addTextChangedListener(new TextWatcher() {

            @Override
            public void onTextChanged(CharSequence s, int start, int before,
                        int count) {

                Toast.makeText(MainActivity.this, edittext_num.getText().toString(),
                            Toast.LENGTH_LONG).show();
            }

            @Override
            public void beforeTextChanged(CharSequence s, int start, int count,
                        int after) {

            }

            public void afterTextChanged(Editable s) {

            }
        });
```

5.8　CheckBox 多项选择

5.8.1　CheckBox 类的结构

多项选择 CheckBox 组件也被称为复选框，该组件常用于某选项的打开或者关闭。它的层次关系如下：

```
java.lang.Object
android.view.View
  android.widget.TextView
    android.widget.Button
      android.widget.CompoundButton
        android.widget.CheckBox
```

5.8.2 CheckBox 类常用的方法

CheckBox 类的常用方法如表 5-10 所示。

表 5-10 CheckBox 类常用的方法

主要方法	功能描述	返回值
dispatchPopulateAccessibilityEvent	在子视图创建时,分派一个辅助事件	boolean(true:完成辅助事件分发 false:没有完成辅助事件分发)
isChecked	判断组件状态是否勾选	boolean(true:被勾选,false:未被勾选)
onRestoreInstanceState	设置视图恢复以前的状态	void
performClick	执行 click 动作,该动作会触发事件监听器	boolean(true:调用事件监听器,false:没有调用事件监听器)
setButtonDrawable	根据 Drawable 对象设置组件的背景	void
setChecked	设置组件的状态	void
setOnCheckedChangeListener	设置事件监听器	void
tooggle	改变按钮当前的状态	void
onCreateDrawableState	为当前视图生成新的 Drawable 状态	int[]

5.8.3 CheckBox 属性

CheckBox 类常用的属性见表 5-11。

表 5-11 CheckBox 类常用的属性

属性名称	描述
android:adjustViewBounds	设置是否保持宽高比,true 或 false
android:cropToPadding	是否截取指定区域用空白代替。单独设置无效果,需要与 scrollY 一起使用。true 或者 false
android:maxHeight	设置图片按钮的最大高度
android:maxWidth	设置图片的最大宽度
android:scaleType	设置图片的填充方式
android:src	设置图片按钮的 drawable
android:tint	设置图片为渲染颜色
android:hint	设置文本为空是所显示的字符

5.8.4 CheckBox 的使用

下面是一个使用 ChekBox 的实例,其效果图见图 5-7。

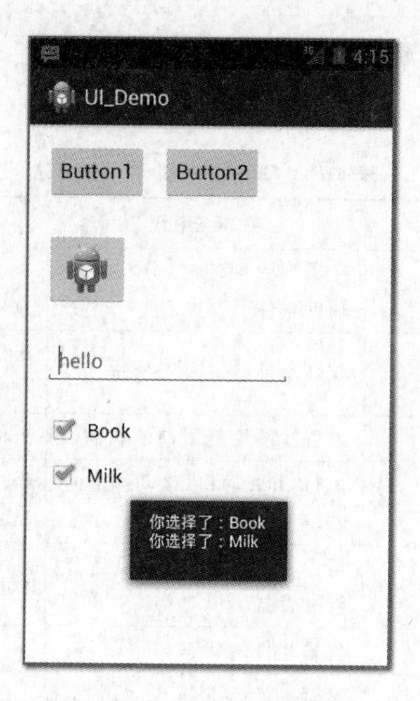

图 5-7　CheckBox 样例效果图

```
<CheckBox
        android:id = "@ + id/checkBox1"
        android:layout_width = "wrap_content"
        android:layout_height = "wrap_content"
        android:layout_alignLeft = "@ + id/editText1"
        android:layout_below = "@ + id/editText1"
        android:layout_marginTop = "20dp"
        android:text = "Book" />

    <CheckBox
        android:id = "@ + id/checkBox2"
        android:layout_width = "wrap_content"
        android:layout_height = "wrap_content"
        android:layout_alignLeft = "@ + id/checkBox1"
        android:layout_below = "@ + id/checkBox1"
        android:text = "Milk" />
```

CheckBox 的使用核心代码：

```
    m_CheckBox1 = (CheckBox) findViewById(R.id.checkBox1);
    m_CheckBox2 = (CheckBox) findViewById(R.id.checkBox2);
    m_CheckBox1.setOnCheckedChangeListener(ocl);
    m_CheckBox2.setOnCheckedChangeListener(ocl);
}

OnCheckedChangeListener ocl = new CheckBox.OnCheckedChangeListener() {
    @Override
```

```
    public void onCheckedChanged(CompoundButton buttonView,
                boolean isChecked) {
        if (isChecked) {
                String s = "";
                if (m_CheckBox1.isChecked()) {
                    s = s + ("你选择了：" + m_CheckBox1.getText()) + "\n";
                }
                if (m_CheckBox2.isChecked()) {
                    s = s + ("你选择了：" + m_CheckBox2.getText()) + "\n";
                }
                Toast.makeText(MainActivity.this, s, Toast.LENGTH_LONG).show();
        }
    }
};
```

5.9 RadioGroup、RadioButton 单项选择

RadioButton 指的是一个单选按钮，它有选中和不选中两种状态，而 RadioGroup 组件也被称为单项按钮组，它可以有多个 RadioButton。一个单选按钮组只可以选中一个按钮，当选择一个按钮时，会取消按钮组中其他已经选中的按钮的选中状态。

5.9.1 类的层次关系

RadioButton 的类层次关系如下：

```
java.lang.Object
android.view.View
    └android.widget.TextView
      └android.widget.Button
         └android.widget.CompoundButton
            └android.widget.RadioButton
```

而 RadioGroup 类的层次关系如下：

```
java.lang.Object
  android.view.View
    android.view.ViewGroup
      android.widget.LinearLayout
        android.widget.RadioGroup
```

5.9.2 RadioGroup 类常用的方法

RadioGroup 中使用到的公共方法如表 5-12 所示。

5.9.3 RadioButton 和 RadioGroup 的综合使用

在 XML 布局中，效果图见图 5-8。

表 5-12　RadioGroup 常用的方法

主 要 方 法	功 能 描 述	返　回　值
addView （View child, int index, ViewGroup. LayoutParams params)	使用指定的布局参数添加一个子视图	void
check (int id)	如果传递－1 作为指定的选择标识符来清除单选按钮组的勾选状态，相当于调用 clearCheck( ) 操作	void
clearCheck ()	清除当前的选择状态，当选择状态被清除，则单选按钮组中的所有单选按钮将取消勾选状态	void
generateLayoutParams (AttributeSet attrs)	基于提供的属性集合返回一个新的布局参数集合	RadioGroup. LayoutParams
getCheckedRadioButtonId ()	返回该单选按钮组中所选择的单选按钮的标识 ID，如果没有勾选，则返回－1	int
setOnCheckedChangeListener (RadioGroup. OnCheckedChangeListener listener)	注册一个当该单选按钮组中的单选按钮勾选状态发生改变时所要调用的回调函数	void
setOnHierarchyChangeListener (ViewGroup. OnHierarchyChangeListener listener)	注册一个当子内容添加到该视图或者从该视图中移除时所要调用的回调函数	void

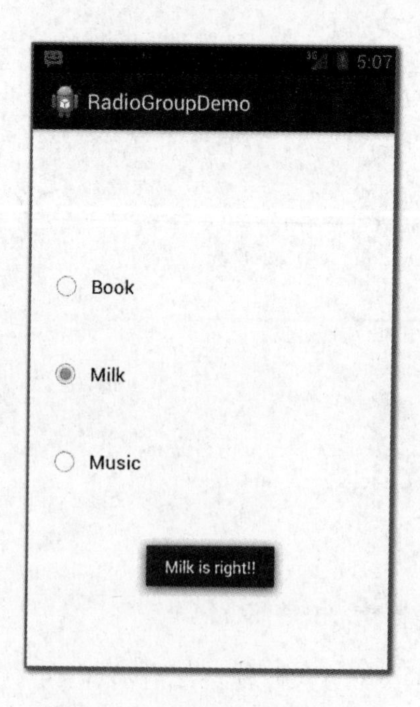

图 5-8　RadioGroup 样例效果图

```
<RelativeLayout xmlns:android = "http://schemas.android.com/apk/res/android"
    xmlns:tools = "http://schemas.android.com/tools"
    android:layout_width = "match_parent"
    android:layout_height = "match_parent"
    android:paddingBottom = "@dimen/activity_vertical_margin"
    android:paddingLeft = "@dimen/activity_horizontal_margin"
    android:paddingRight = "@dimen/activity_horizontal_margin"
    android:paddingTop = "@dimen/activity_vertical_margin"
    tools:context = ".MainActivity" >

    <RadioGroup
        android:id = "@ + id/RadioGroup01"
        android:layout_width = "wrap_content"
        android:layout_height = "wrap_content"
        android:layout_alignParentLeft = "true"
        android:layout_alignParentTop = "true"
        android:orientation = "vertical" >

        <RadioButton
            android:id = "@ + id/radioButton1"
            android:layout_width = "wrap_content"
            android:layout_height = "wrap_content"
            android:layout_marginTop = "98dp"
            android:text = "Book" />

        <RadioButton
            android:id = "@ + id/radioButton2"
            android:layout_width = "wrap_content"
            android:layout_height = "wrap_content"
            android:layout_marginTop = "36dp"
            android:text = "Milk" />

        <RadioButton
            android:id = "@ + id/radioButton3"
            android:layout_width = "wrap_content"
            android:layout_height = "wrap_content"
            android:layout_marginTop = "36dp"
            android:text = "Music" />
    </RadioGroup>

</RelativeLayout>
```

核心代码如下：

```
protected void onCreate(Bundle savedInstanceState) {
        super.onCreate(savedInstanceState);
        setContentView(R.layout.activity_main);
        m_RadioGroup = (RadioGroup) findViewById(R.id.RadioGroup01);
        m_Radio1 = (RadioButton) findViewById(R.id.radioButton1);
```

```
            m_Radio2 = (RadioButton) findViewById(R.id.radioButton2);
            m_Radio3 = (RadioButton) findViewById(R.id.radioButton3);
            m_RadioGroup
                    .setOnCheckedChangeListener(new RadioGroup.OnCheckedChangeListener() {
                        @Override
                        public void onCheckedChanged(RadioGroup group, int checkedId) {
                            if (checkedId == m_Radio2.getId())

                                Toast.makeText(MainActivity.this,
                                        m_Radio2.getText() + " is right!!",
            Toast.LENGTH_LONG).show();
                        }
                    });
        }

    }
```

5.10 Spinner 下拉列表

5.10.1 Spinner 类的层次关系

Spinner 功能类似 RadioGroup，相比 RadioGroup，Spinner 提供了体验性更强的 UI 设计模式。一个 Spinner 对象包含多个子项，每个子项只有两种状态：选中或未选中。Spinner 类的层次关系如下：

```
java.lang.Object
android.view.View
  android.view.ViewGroup
    android.widget.AdapterView<T extends android.widget.Adapter>
        android.widget.AbsSpinner
           android.widget.Spinner
```

5.10.2 Spinner 类的主要方法

Spinner 类的主要方法如表 5-13 所示。

表 5-13 Spinner 常用的方法

主要方法	功能描述
public int getBaseline()	返回这个控件文本基线的偏移量。如果这个控件不支持基线对齐，那么方法返回－1 返回值：返回控件基线左边边界位置，不支持时返回－1
public CharSequence getPrompt()	当对话框弹出的时候显示的提示（即：获得弹出视图上的标题字）

续表

主 要 方 法	功 能 描 述
public void onClick (DialogInterface dialog, int which)	当单击弹出框中的项时这个方法将被调用
public boolean performClick()	如果它被定义就调用此视图的 OnClickListener 返回值为 True 一个指定的 OnClickListener 被调用,为 False 时不被调用
public void setOnItemClickListener (OnItemClickListener l)	Spinner 不支持 item 的单击事件,调用此方法将引发异常
public void setPromptId(CharSequence prompt)	设置对话框弹出的时候显示的提示(弹出视图上的标题字)
public void setPromptId(int promptId)	设置对话框弹出的时候显示的提示 参数:prompted 当对话框显示是显示这个资源 id 所代表的提示
protected void onDetachedFromWindow ()	当这个视图从屏幕上卸载时候被调用。在这一点上不再绘制视图
protected void onLayout (boolean changed, int l, int t, int r, int b)	当 View 要为所有子对象分配大小和位置时,调用此方法。派生类及其子项们应该重载这个方法和调用布局每一个子项

5.10.3 Spinner 的使用示例

首先在 xml 中声明 Spinner,这里同时声明了一个 TextView 用于显示 Spinner 的监听结果(见图 5-9):

```
<TextView android:id="@+id/TextView01"
        android:layout_width="fill_parent"
        android:layout_height="wrap_content"
        android:text="@string/hello"/>
<Spinner android:id="@+id/Spinner01"
        android:layout_width="300dip"
        android:layout_height="wrap_content">
</Spinner>
```

然后就可以在 Activity 中使用了:

```
final TextView textview = (TextView)findViewById(R.id.TextView01);
        Spinner spinner = (Spinner) findViewById(R.id.Spinner01);
        final List<String> list = new ArrayList<String>();
        list .add("Spinner 子项 1");
        list .add("Spinner 子项 2");
        list .add("Spinner 子项 3");
        //将可选内容 list 与 ArrayAdapter 相连接
        ArrayAdapter<String> adapter = new ArrayAdapter<String>(this, android.R.layout.
simple_spinner_item, list);
        //设置下拉列表的风格
```

```
        adapter.setDropDownViewResource(android.R.layout.simple_spinner_dropdown_item);
       //将 Adapter 添加到 Spinner
        spinner.setAdapter(adapter);
```


完成了 Spinner 的显示代码后,接下来就是添加事件监听了。

```
//添加事件监听
spinner.setOnItemSelectedListener(new Spinner.OnItemSelectedListener(){
                @Override
                public void onItemSelected(AdapterView<?> arg0, View arg1,
                        int arg2, long arg3) {
                    textview.setText("你当前选择的是: " + list.get(arg2));
                }
                @Override
                public void onNothingSelected(AdapterView<?> arg0) {

                }
            });
```

示例的使用效果图如图 5-9 所示。

图 5-9 Spinner 样例效果图

5.11 RatingBar 下拉列表

5.11.1 RatingBar 类的层次关系

RatingBar 是基于 SeekBar 和 ProgressBar 的扩展,用星形来显示等级评定。使用 RatingBar 的默认大小时,用户可以触摸/拖动或使用键来设置评分,它有两种样式(小风格用 ratingBarStyleSmall,大风格用 ratingBarStyleIndicator),其中大的只适合指示,不适合于用户交互。

当使用可以支持用户交互的 RatingBar 时，无论将控件（widgets）放在它的左边还是右边都是不合适的。

只有当布局的宽被设置为 wrap content 时，设置的星形数量（通过函数 setNumStars (int)或者在 XML 的布局文件中定义）将显示出来（如果设置为另一种布局宽的话，后果无法预知）。次级进度一般不应该被修改，因为它仅仅是被当作星形部分内部的填充背景。

RatingBar 的 XML 属性见表 5-14。

表 5-14 RatingBar 常用的属性

属性名称	功能描述
android:isIndicator	RatingBar 是否是一个指示器（用户无法进行更改）
android:numStars	显示的星形数量，必须是一个整型值，如"100"
android:rating	默认的评分，必须是浮点类型，如"1.2"

RatingBar 的层次关系如下所示：

```
java.lang.Object
        android.view.View
                android.widget.ProgressBar
                        android.widget.AbsSeekBar
                                android.widget.RatingBar
```

5.11.2 RatingBar 类的主要方法

RatingBar 类的主要方法如表 5-15 所示。

表 5-15 RatingBar 常用的方法

主要方法	功能描述
public int getNumStars()	返回显示的星形数量
public RatingBar. OnRatingBarChangeListener getOnRatingBarChangeListener()	监听器（可能为空）监听评分改变事件
public float getRating()	获取当前的评分（填充的星形的数量）
public float getStepSize()	获取评分条的步长
public boolean isIndicator()	判断当前的评分条是否仅仅是一个指示器（注：即能否被修改）
public void setIsIndicator (boolean isIndicator)	设置当前的评分条是否仅仅是一个指示器（这样用户就不能进行修改操作了）
public synchronized void setMax (int max)	设置评分等级的范围，从 0～max
public void setNumStars (int numStars)	设置显示的星形的数量
public void setOnRatingBarChangeListener (RatingBar .OnRatingBarChangeListener listener)	设置当评分等级发生改变时回调的监听器
public void setRating (float rating)	设置分数（星形的数量）
public void setStepSize (float stepSize)	设置当前评分条的步长（step size）

5.11.3 RatingBar 的使用示例

在 XML 中声明三种样式的 RatingBar：

```
<RatingBar android:layout_width = "wrap_content"
        android:layout_height = "wrap_content" style = "?android:attr/ratingBarStyleIndicator"
        android:id = "@ + id/ratingbar_Indicator" />
    <RatingBar android:layout_width = "wrap_content"
        android:layout_height = "wrap_content" style = "?android:attr/ratingBarStyleSmall"
        android:id = "@ + id/ratingbar_Small" android:numStars = "20" />
    <RatingBar android:layout_width = "wrap_content"
        android:layout_height = "wrap_content" style = "?android:attr/ratingBarStyle"
        android:id = "@ + id/ratingbar_default" />
```

在 Activity 中声明并显示：

```
final RatingBar ratingBar_Small = (RatingBar)findViewById(R.id.ratingbar_Small);
        final RatingBar ratingBar_Indicator = (RatingBar)findViewById(R.id.ratingbar_
Indicator);
        final RatingBar ratingBar_default = (RatingBar)findViewById(R.id.ratingbar_default);

        ratingBar_default.setOnRatingBarChangeListener(new RatingBar.OnRatingBarChangeListener(){

    public void onRatingChanged(RatingBar ratingBar, float rating,
      boolean fromUser) {
     ratingBar_Small.setRating(rating);
     ratingBar_Indicator.setRating(rating);
     Toast.makeText(RatingBarActivity.this, "rating:" + String.valueOf(rating),
       Toast.LENGTH_LONG).show();
    }});
```

示例的使用效果图如图 5-10 所示。

图 5-10　RatingBar 样例图

5.12　本 章 小 结

通过本章的学习，我们已经掌握了 Android 常见的 UI 控件。

5.13　习题与课外阅读

5.13.1　习题

请设计出如图 5-11 所示的效果用户界面。

图 5-11　习题样例

5.13.2　课外阅读

访问下列技术网站，了解一下 Android 本书未提及的其他 UI 控件的属性和使用：
http://developer.android.com/training/index.html

第6章 Android UI 线程通信

Android UI 线程通信机制比较复杂，这部分内容是 Android 程序设计的难点，是 Android UI 程序与 PC Java GUI 程序设计最大的不同点，也是 Android 的特色内容之一。许多资深的 PC Java 程序员，最初接触 Android UI 设计时，由于沿用了 PC Java 应用程序 GUI 设计的思维模式，编写的 Android 程序无任何错误，可以编译打包安装到 Android 手机上，但是一旦运行程序，程序立刻崩溃，这是非常令人沮丧的事情，其原因主要是不了解 Android UI 线程通信机制。

Android 的 UI 操作并不是线程安全的，Android 将涉及 UI 的操作限定在 UI 线程中完成，任何在 UI 线程之外的线程操作 UI 或阻塞 UI 线程的代码将导致程序异常退出。对 Android UI 操作通常采用子线程进行异步处理的技术方案，主要利用 Handler 或 AsyncTask 配合主线程异步更新 UI 界面。

本章的学习目标：

- 掌握 Android UI 操作与线程概念；
- 掌握使用 Handler 更新 Android UI 的方法；
- 掌握使用 AsyncTask 更新 Android UI 的方法。

6.1 Android UI 操作与线程

基于计算机操作系统知识，我们可以知道一个进程中可以同时有多个线程在运行，而这些线程可以同时运行同一段代码，如果每次运行的结果和单线程执行的结果相同，而且其他的变量的值也和预期的是一样的，就是线程安全的，否则不是线程安全的。线程不安全，可以通过加锁等方法消除。

Android 的 UI 操作并不是线程安全的，主要为了减少加锁带来的性能开销。Android 将涉及 UI 的操作限定在 UI 线程中完成，任何在 UI 线程之外的线程操作 UI 将导致程序异常退出。通常情况下，当一个应用程序第一次启动时，就会同时启动一个 UI 线程，即主线程，在这个线程中刷新 UI 是安全的。

但是并非所有的操作都可以在主线程中完成，例如，联网读取数据，或者读取本地较大的一个文件的时候，如果将这些操作放在主线程中的，会阻塞主线程导致应用程序无响应(Application Not Response)或程序闪退，因此不能把这些操作放在主线程中。

这就要求开发者必须遵循两条法则：

(1) 不能阻塞 UI 线程；

(2) 确保只在 UI 线程中访问 Android UI。

于是,开启子线程进行异步处理的技术方案应运而生。通常采用的技术是:利用 Handler 或 AsyncTask 配合主线程异步更新 UI 界面,从而实现了不阻塞主线程(UI 线程),且 UI 的更新只能在主线程中完成。

6.2 相关概念

(1) Thread:线程是一个并发的执行单位(A Thread is a concurrent unit of execution),线程只能在进程中执行,一个进程可包含多个线程。

(2) Message:消息,其中包含了消息 ID、消息处理的数据等。Message 实例对象的创建或取得可以有以下几种常用的方式:

- Message 类中的静态方法 obtain()获得方法有多个重载版本可供选择。作用是从 Message Pool 中取出一个 Message,如果 Message Pool 中已经没有 Message 可取则新建一个 Message 返回,同时用对应的参数给得到的 Message 对象赋值。
- 通过 Handlcr 对象的 obtainMessage()获取 个 Message 实例。
- 如果 Message Pool 中没有可用的 Message 实例,则可以创建一个 Message 对象。调用 removeMessages()时,将 Message 从 MessageQueue 中删除,同时放入到 Message Pool 中。

(3) MessageQueue:消息队列,每一个线程最多只可以拥有一个 MessageQueue 数据结构。用来存放 Handler 发送过来的消息,并按照 FIFO 规则执行。当然,存放 Message 并非实际意义上的保存,而是将 Message 以链表的方式串联起来的,等待 Looper 的抽取。创建一个线程的时候,并不会自动创建其 MessageQueue。

(4) Looper:是 MessageQueue 的管理者。不断地从 MessageQueue 中抽取 Message 执行。因此,一个 MessageQueue 需要一个 Looper。每一个 MessageQueue 都不能脱离 Looper 而存在,Looper 对象的创建是通过 prepare()函数来实现的。同时每一个 Looper 对象和一个线程关联。通过调用 Looper. myLooper()可以获得当前线程的 Looper 对象。Looper 和 MessageQueue 一一对应。创建一个 Looper 对象时,会同时创建一个 MessageQueue 对象。除了主线程有默认的 Looper,其他线程默认是没有 MessageQueue 对象的,所以,不能接受 Message。如需要接受,自己定义一个 Looper 对象(通过 prepare()函数),这样该线程就有了自己的 Looper 对象和 MessageQueue 数据结构了。创建主线程时,会创建一个默认的 Looper 对象,而创建 Looper 对象时,将自动创建一个 MessageQueue。其他非主线程,不会自动创建 Looper,需要 Looper 时,通过调用 prepare()函数来实现。

(5) Handler:消息处理者,负责 Message 的发送及处理。使用 Handler 时,需要实现 handleMessage(Message msg)方法来对特定的 Message 进行处理,例如更新 UI 等。将消息传递给 Looper,这是通过 Handler 对象的 sendMessage()来实现的。继而由 Looper 将 Message 放入 MessageQueue 中。当 Looper 对象看到 MessageQueue 中含有 Message,就将其抽取出去。该 handler 对象收到该消息后,调用相应的 handleMessage()方法对其进行处理。

由此可见,一个 Message 经由 Handler 的发送、MessageQueue 的入队、Looper 的抽取,又再一次地回到 Handler 的怀抱。而绕的这一圈,也正好帮助我们将同步操作变成了异步操作,如图 6-1 所示。

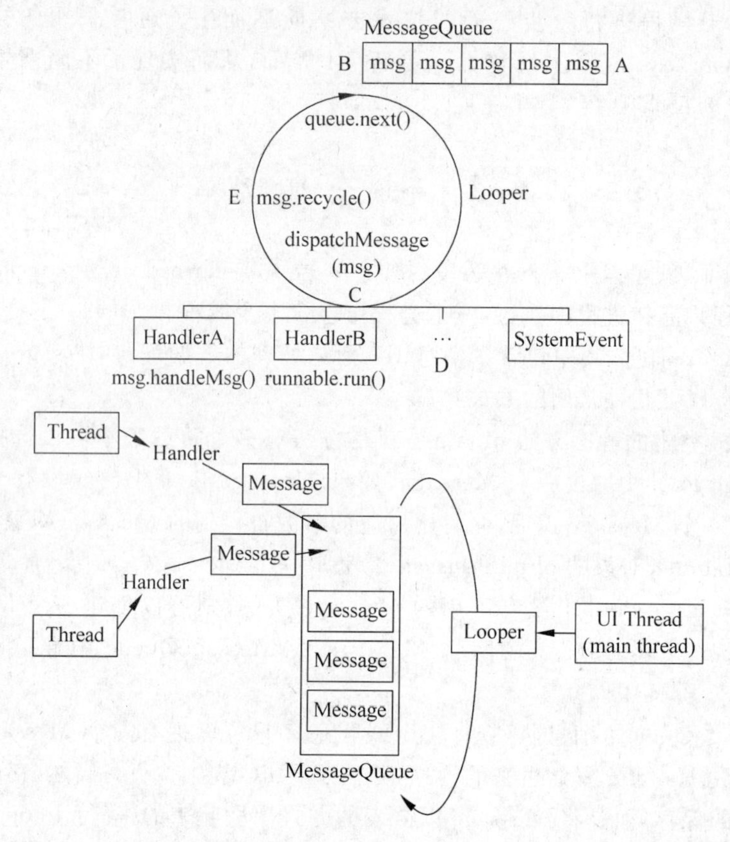

图 6-1 Thread、Handler、Message MessageQueue、Looper 之间的关系图

6.3 Handler 的使用

Handler 负责维护两个队列。Handler 可以分发 Message 消息对象和 Runnable 对象(任务对象)到主线程中。所以每个 Handler 对象都可以维护两个类型队列：Message 队列和 Runnable 队列，这两个队列来分别是：

(1) 发送、接受、处理消息——Message 队列(消息队列)；

(2) 启动、结束、休眠线程——Runnable 队列(任务队列)。

6.3.1 Handler 处理 Message 队列

Handler 对象维护一个 MessageQueue，有新的 Message 通过 sendMessage()送来的时候，把它放在队尾，之后排队到处理该消息的时候，由主线程的 Handler 对象 handleMessage()方法处理。整个过程也是异步的。向队列添加消息常用方法有：

- handler. sendMessage(Message)——向队列添加 Message。
- handler. sendMessageDelayed(Message，long)——延迟一定时间后，将消息发送到消息队列。
- handler. sendMessageAtTime(Message，long)——定时将消息发送到消息队列。

消息的具体处理过程，需要在 new Handler 对象时使用匿名内部类重写 Handler 的

handleMessage(Message msg)方法,在该方法中处理。

【例 6-1】 编写一个 Handler 维护的消息队列例子,利用子线程与主线程通信,更新 UI 界面。子线程请求网络数据,获取网络上的图片;当子线程完成络图片获取后,发送消息给主线程,通知主线程;主线程获得消息后,利用子线程获取的图片来更新 UI 界面。

```
//创建线程
Runnable runnable = new Runnable() {
    @Override
    public void run() {
        // TODO Auto - generated method stub
        HttpClient httpClient = new DefaultHttpClient();
        HttpGet httpGet = new HttpGet(
    "http://cms.csdnimg.cn/article/201303/25/514fb45ab4a82_middle.jpg");
        final Bitmap bitmap;
        try {
            HttpResponse httpResponse = httpClient.execute(httpGet);
            bitmap = BitmapFactory.decodeStream(httpResponse.getEntity()
                    .getContent());
            Message msg = new Message();
            msg.what = MSG_SUCCESS;
            msg.obj = bitmap;
            mHandler.sendMessage(msg);
            // 另一种写法
            // Message msg = mHandler.obtainMessage();
            // msg.what = MSG_SUCCESS;
            // msg.obj = bitmap;
            // msg.sendToTarget();
        } catch (ClientProtocolException e) {
            // TODO Auto - generated catch block
            e.printStackTrace();
            mHandler.sendEmptyMessage(MSG_FAILURE);
        } catch (IOException e) {
            // TODO Auto - generated catch block
            e.printStackTrace();
            mHandler.sendEmptyMessage(MSG_FAILURE);
        }
    }
};

            //开启线程,发送消息
            Thread mThread = new Thread(runnable);
            mThread.start();

//处理消息
    private Handler mHandler = new Handler() {
        @Override
        public void handleMessage(Message msg) {
            // TODO Auto - generated method stub
            switch (msg.what) {
```

```
                case MSG_SUCCESS:
                        mImageView.setImageBitmap((Bitmap) msg.obj);
                        Toast.makeText(getApplication(), "获取图片成功", Toast.LENGTH_LONG)
                                        .show();
                        break;
                case MSG_FAILURE:
                        Toast.makeText(getApplication(), "获取图片失败", Toast.LENGTH_LONG)
                                        .show();
                        break;
                }
        }
};
```

6.3.2 Handler 处理 Runnable 队列

Handler 对象维护一个线程队列，当有新的 Runnable 送来 post()的时候，把它放在队尾，而处理 Runnable 的时候，Handler 从队头取出 Runnable 执行。

向队列添加线程常用方法：

- handler.post(Runnable)——将 Runnable 直接添加入队列。
- handler.postDelayed(Runnable, long)——延迟一定时间后，将 Runnable 添加入队列。
- handler.postAtTime(Runnable,long)——定时将 Runnable 添加入队列。

终止线程常用方法：

handler.removeCallbacks(Runnable r)——将 Runnable 从 Runnable 队列中取出，使线程停止执行。

值得注意的是，handler.post()方法并未真正新建线程，只是在原线程上执行而已，即还是运行当前 handler 创建的线程(一般是位于主线程)中，一些耗时的操作，还是不能在此线程中的 run 方法中实现。Handler 只是调用了 Runnable 对象的 run 方法。

【例 6-2】 演示 handler.post(Runnable r)方法，处理线程队列，设置某个时刻在主线程中执行 Runnable 任务。

```
    // 创建一个 Handler 对象
    Handler updateBarHandler = new Handler() {
        @Override
        public void handleMessage(Message msg) {
        // handler 收到消息后更新 UI 界面，并继续把此线程加入到主线程队列执行

            progressBar.setProgress(msg.arg1);
            updateBarHandler.post(updateBarThread);
        }

    };

// 将一个线程加入到主线程队列
    updateBarHandler.post(updateBarThread);
```

```
// 更新 ProgressBar 的线程对象
Runnable updateBarThread = new Runnable() {
    int i = 0;

    @Override
    public void run() {
        // 从打印的日志可以看到此线程还是属于主线程
        i = i + 10;
        Message msg = updateBarHandler.obtainMessage();
        msg.arg1 = i;
        try {
            Thread.sleep(1000);
        } catch (InterruptedException e) {
            e.printStackTrace();
        }
        textView.setText("当期的进度值为: " + msg.arg1);
        if (i <= 100) {
            updateBarHandler.sendMessage(msg);
        } else {
            updateBarHandler.removeCallbacks(updateBarThread);
        }
    }
};

// 移除此线程
updateBarHandler.removeCallbacks(updateBarThread);
```

从 Handler 维护的两个队列,可以总结出它的两个作用:

(1) 安排(定时执行或者也叫作延迟执行)消息或 Runnable 在某个主线程中某个地方执行(使用 POST 方法),注意这个 Runnable 并非是子线程,而是运行在主线程中的,这个延迟操作一般常用于应用程序的欢迎界面的跳转等。

(2) 安排一个动作在另外的线程中执行,配合主线程更新 UI。

6.4 子线程和主线程的双向通信

6.3 节介绍了 Android 的线程中的单向通信,本节介绍 Android 中子线程和主线程的双向通信,首先来介绍 Looper。

6.4.1 Looper 介绍

Looper 是 MessageQueue 的管理者。每一个 MessageQueue 都不能脱离 Looper 而存在,Looper 对象的创建是通过 prepare()函数来实现的。同时每一个 Looper 对象和一个线程关联。通过调用 Looper. myLooper()可以获得当前线程的 Looper 对象,创建一个

Looper 对象时，会同时创建一个 MessageQueue 对象。

- 除了主线程(系统会自动为其创建 Looper 对象，开启消息循环)有默认的 Looper，其他线程默认是没有 MessageQueue 对象的，也没有消息循环的机制。所以，不能接收 Message。
- 如需要在子线程接收消息，需要自己定义一个 Looper 对象(通过 prepare()函数)，这样该线程就有了自己的 Looper 对象和 MessageQueue 数据结构了。Looper 从 MessageQueue 中取出 Message 然后，交由 Handler 的 handleMessage()方法进行处理。

常用方法如下：

Looper. prepare()——启用 Looper。

Looper. loop()——让 Looper 开始工作，从消息队列里取消息。

6.4.2 Looper 使用的注意事项

(1) Looper 类用来为一个线程开启一个消息循环。一个线程只能有一个 Looper，对应一个 MessageQueue。

(2) 通常是通过 Handler 对象来与 Looper 进行交互的。Handler 可看作是 Looper 的一个接口，用来向指定的 Looper 发送消息及定义处理方法。默认情况下 Handler 会与其被定义时所在线程的 Looper 绑定，比如，Handler 在主线程中定义，那么它是与主线程的 Looper 绑定。mainHandler = new Handler()等价于 new Handler(Looper. myLooper())。Looper. myLooper()用于获取当前线程的 Looper 对象。类似地，Looper. getMainLooper()用于获取主线程的 Looper 对象。

(3) 在子线程中直接 new Handler()会报错误"java. lang. RuntimeException: Can't create handler inside thread that has not called Looper. prepare()"，原因是非主线程中默认没有创建 Looper 对象，需要先调用 Looper. prepare()启用 Looper。

(4) Looper. loop()让 Looper 开始工作，从消息队列里取消息，处理消息。注意：写在 Looper. loop()之后的代码不会被执行，这个函数内部应该是一个循环，当调用 mHandler. getLooper(). quit()后，Loop 函数才会中止，其后的代码才能得以运行。

(5) 在创建 Handler 之前，为该线程准备好一个 Looper(Looper. prepare)，然后让这个 Looper 运行起来(Looper. loop)，抽取 Message，这样，Handler 才能正常工作。因此，Handler 处理消息总是在创建 Handler 的线程中运行。而在消息处理中，不乏更新 UI 的操作，不正确的线程直接更新 UI 将引发异常。因此，需要搞清楚 Handler 是在哪个线程中创建的。

(6) 一个 Looper 只有处理完一条 Message 才会读取下一条，所以消息的处理是阻塞形式的(handleMessage()方法里不应该有耗时操作，可以将耗时操作放在其他线程执行，操作完后发送 Message(通过 sendMessges 方法)，然后由 handleMessage()更新 UI)。

【例 6-3】 主线程向子线程发送消息，子线程接到消息后加入自己的数据，又把消息返回主线程。

```
//mainHandler 是在主线程创建的,主线程自带 Looper,无须创建 looper
        mainHandler = new Handler(){
                @Override
                public void handleMessage(Message msg) {
                        super.handleMessage(msg);
                        switch(msg.what) {
                        case TOMAIN:
                                main_txt.setText(msg.obj.toString());
                                break ;
                        }
                }

        } ;

    //创建子线程,并开启它
        LooperThread looperThread = new LooperThread();
        looperThread.start();
//子线程实现
  class LooperThread extends Thread {

        public void run() {
                // 创建该线程的 Looper 对象,用于接收消息
                //在非主线程中是没有 Looper 的,所以在创建 Handler 前一定要使用 prepare()创建
                //一个 Looper
                Looper.prepare();
                //childHandler 是在子线程中创建的
                childHandler = new Handler() {
                        public void handleMessage(Message msg) {
                            switch(msg.what) {
                            case TOCHILD: // 子线程接收主线程发送来的消息
                                    Message mainMsg = mainHandler.obtainMessage() ;
                                    mainMsg.what = TOMAIN ;
                                    mainMsg.obj = msg.obj + "\n\n 这是子线程加入的字符串,又发
  向主线程!!!";
                                  mainHandler.sendMessage(mainMsg) ;
                                  break ;
                            }
                        }
                   };
                   // 创建消息队列
                   Looper.loop();
        }
                   //创建发向子线程的消息
                        Message childMsg = childHandler.obtainMessage();
                        childMsg.what = TOCHILD;
                        childMsg.obj = "这是从主线程发向子线程的字符串!!!";
                        childHandler.sendMessage(childMsg);
```

6.5 AsyncTask 异步任务类

6.5.1 AsyncTask 简介

为了更加方便编写后台线程与 UI 线程交互程序，自 Android SDK 1.5 之后推出了 AsyncTask 类。AsyncTask 是一个抽象类，它的内部其实也是结合了 Thread 和 Handler 来实现异步线程操作，但是它形成了一个通用线程框架，更清晰简单。AsyncTask 运行在用户界面中，执行异步操作，并且把执行结果发布在 UI 线程上，且也不需要处理线程和 Handler。使用 AsyncTask 类，要明确“三个参数，五个回调方法，四个注意事项”。

6.5.2 AsyncTask 的三个参数

AsyncTask 异步工作的参数与返回值被泛型的三个参数为：

(1) Params，启动任务执行的输入参数，比如，HTTP 请求的 URL。该参数传递给后台任务的(doInBackground)参数，需要注意的是，该参数可以为可变长输入参数(其实就是数组)，即：可以一次性启动执行多个下载任务列表，例如传入一组 String[]。

(2) Progress，后台计算执行过程中，进度单位(progress units)的类型(就是后台程序已经执行了百分之几了)，一般是 Integer 或者 Double(注意是封装类而不是原始的 int 或 double)。

(3) Result，后台任务执行完成后返回的结果的类型，也是 onPostExecute(Result)的传入参数类型，二者必须一致。

AsyncTask 不一定总是需要使用上面的全部三个参数。如果不使用上面的参数，只需将参数定义为 void 类型即可。三个参数皆可为空，只需要传入 AsyncTask＜Void，Void，Void＞即可。

6.5.3 AsyncTask 的五个回调方法

AsyncTask 运行的五个回调方法对应五个状态，如表 6-1 所示。

表 6-1 AsyncTask 后台线程运行的五个回调方法

状 态 名 称	说 明
1. 准备运行	准备运行：onPreExecute()，该回调函数在任务被执行之后立即由 UI 线程调用。这个步骤通常用来建立任务，在用户接口(UI)上显示进度条
2. 正在后台运行	正在后台运行：doInBackground(Params...)，该回调函数由后台线程在 onPreExecute()方法执行结束后立即调用。通常在这里执行耗时的后台计算。计算的结果必须由该函数返回，并被传递到 onPostExecute()中。在该函数内也可以使用 publishProgress(Progress...)来发布一个或多个进度单位(unitsof progress)。这些值将会在 onProgressUpdate(Progress...)中被发布到 UI 线程
3. 进度状态更新	进度更新：onProgressUpdate(Progress...)，该函数由 UI 线程在 publishProgress(Progress...)方法调用完后被调用。一般用于动态地显示一个进度条
4. 完成后台任务	完成后台任务：onPostExecute(Result)，当后台计算结束后调用。后台计算的结果会被作为参数传递给这一函数被 UI thread 调用，后台的计算结果将通过该方法传递到 UI thread
5. 取消任务	取消任务：onCancelled ()，在调用 AsyncTask 的 cancel()方法时调用

6.5.4 AsyncTask 使用的四点注意事项

(1) AsyncTask 必须声明在 UI 线程上；

(2) AsyncTask 必须在 UI 线程上实例化；

(3) execute()方法必须在 UI 线程中调用；

(4) 不要手动的调用 onPreExecute()、onPostExecute(Result)、doInBackground(Params...)、onProgressUpdate(Progress...)这几个方法。

需要说明的是：

- AsyncTask 只能被执行一次，若多次调用将会出现异常；
- doInBackground 方法的返回值类型必须与 onPostExecute 方法的参数类型必须一致，这两个参数在 AsyncTask 声明的泛型参数列表中指定。

【例 6-4】 编写一个程序下载网络文件，存放到手机的 SD 卡根目录中，下载完成后，刷新 UI 界面。在下载过程中，即时刷新下载进度条，单击停止按钮，停止界面刷新。

```
/*
 * 异步任务类 MyAsyncTask
 */
// 第一个为 doInBackground 接受的参数，第二个为显示进度的参数，第三个为 doInBackground 返回
// 和 onPostExecute 传入的参数
private class MyAsyncTask extends AsyncTask<String, Integer, String> {

    @Override
    protected String doInBackground(String... params) {
        // 后台线程执行 doInBackground()，不可以在 doInBackground()操作 ui，调用 publishProgress
// 更新进度，这里在 download_Apk()中调用了
        // 此时会调用 onProgressUpdate(Integer...progress)更新进度条(进度用整型数表示，因
// 此 AsyncTask 的第二个模板参数是 Integer)
        download_Apk(params[0]);
        // doInBackground 把返回值传给 onPostExecute()
        return "百度客户端下载成功!";
    }

    @Override
    protected void onCancelled() {
        super.onCancelled();
    }

    @Override
    // 当后台任务执行完成后，调用 onPostExecute(Result)，传入的参数是 doInBackground()中返
// 回的对象，在这里更新 UI 界面
    protected void onPostExecute(String result) {
        result_txt.setText("当前下载状态: " + result);
    }
    @Override
    // 这里是执行线程前，做一些初始化界面的操作
    protected void onPreExecute() {
```

```
            download_bar.setProgress(0);                                    // 进度条复位
            result_txt.setText("当前下载状态：正在下载中.....");
        }
        @Override
        // 在这里实时更新下载进度
        protected void onProgressUpdate(Integer... values) {
            AsyncTaskActivity.this.download_bar.setProgress(values[0]);  // 设置进度
            download_info_txt.setText("当前进度值：" + values[0]);
        }
    }
//开启任务,每创建一个对象,execute 方法只能被调用一次
            myAsyncTask = new MyAsyncTask();
            // 此地址字符串被传递给了 doInBackground 这个方法
            myAsyncTask
        .execute("http://gdown.baidu.com/data/wisegame/378672b311373c82/baidu.apk");
            start_btn.setEnabled(false);
```

6.6 本章小结

通过本章的学习，我们已经掌握了 Android UI 线程通信机制，掌握了采用子线程进行异步处理的技术方案，学会利用 Handler 或 AsyncTask 配合主线程异步更新 UI 界面。

6.7 习题与课外阅读

6.7.1 习题

(1) 开发一个数字钟程序，实现动态显示当前时间。

(2) 开发一个下载程序，从网络上抓取一段文字，显示在 TextView 内。

6.7.2 课外阅读

(1) 阅读 Handler、AsyncTask、Message API 技术问题，加深对相关概念的理解：
http://developer.android.com/training/index.html

(2) 阅读 Android Background Processing with Handlers and AsyncTask and Loaders - Tutorial。

http://www.vogella.com/tutorials/AndroidBackgroundProcessing/article.html

第7章 Intent与组件通信

传统的PC各个不同应用程序之间的功能集成与互操作非常困难,不容易将一个应用与其他已编译好的应用软件的功能集成在一起。Android应用程序与PC应用程序最大的不同点在于:一个Android应用程序中的组件可以与其他应用或者组件通信,实现应用与组件之间的逻辑功能动态集成,动态集成后的组件逻辑功能严密,如同一个逻辑严密PC应用软件一样,其秘诀就在于Android系统的Intent与组件通信机制。Android应用系统中的不同的组件(如Activity、Service、BroadcastReceiver等),都可以使用Intent互相进行通信。如,利用这种机制可以把一个应用程序中的文字或图片,使用Intent直接共享给QQ或微信好友。

通过本章的学习可以让读者理解Android系统中的组件通信原理,掌握使用Intent调用其他组件实现无缝系统集成的方法,掌握监听和发送广播消息的方法。

本章学习目标:

- 了解Intent的概念及使用;
- 掌握Android应用中组件点对点通信方法;
- 掌握Android应用中组件广播与监听方法;
- 掌握使用Intent启动Activity的方法。

7.1 Intent简介

Intent是Android系统中提供的来实现应用间的交互实现系统无缝集成的组件。Intent负责对响应目标组件的组件对象或动作(Action)、类别(Category)以及附加数据(Data)进行描述,Android系统则根据该Intent的描述,负责找到并启动对应的组件,并将Intent传递给调用的组件,并完成组件的调用。Intent可以实现应用程序内部或应用程序之间的交互,在满足一定安全机制下任何组件都可以调用或被其他组件调用,体现出Android应用程序中的组件“人人为我,我为人人”的原则(见图7-1)。

- Intent提供响应目标组件、动作(Action)、类别(Category)以及附加数据(Data)进行描述。
- Android系统据Intent的描述,找到并启动相应组件,并将Intent传递给调用的组件,完成组件的调用。

图7-1 组件通信与Android系统

Intent可以看作是不同组件之间的通信的通道。如果把Activity比作Web服务器上Html网页的话,Intent就像HTTP协议请求的数据报头信息。

Intent 可以用于启动一个 Activity、Service、发送广播(broadcast)消息,实现组件之间的通信(见图 7-2)。

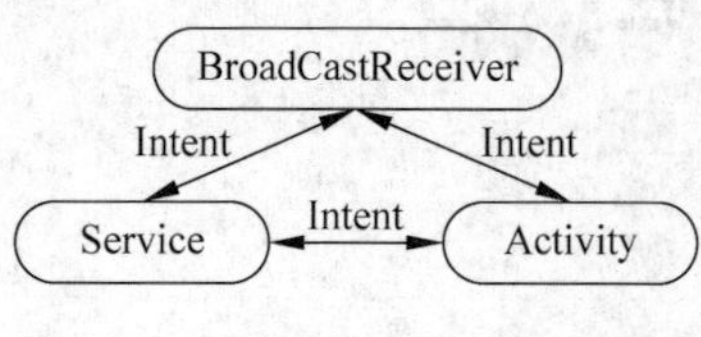

图 7-2　Android 系统四大组件之间的关系

例如,在一个 Activity 中启动另一个 Activity,可以通过调用 startActivity()来实现,或者由一个 Activity 发送广播 Intent( )来传递给所有感兴趣的 Broadcast Receiver,再或者由 startService()/bindservice()来启动一个后台的 service。Intent 主要是用来启动调用其他的 Activity 或者 Service,可以将 Intent 理解成 Activity 或 Service 之间的粘合剂。

7.2　Intent 的构成

Intent 包含的内容有:

(1) 动作(Action)。Action 用来指明要实施的动作。详细内容可查阅 Android SDK API 文档中的 Android. content. intent 类,其中的 constants 定义了所有的 Action,常见的 Activity Action 见表 7-1,标准的 Activity 广播 Action 见表 7-2。

表 7-1　常见的标准的 Activity Actions

Action 名称	功能与作用
ACTION_MAIN	作为一个主要的进入口,不接受数据传入
ACTION_VIEW	向用户显示数据
ACTION_ATTACH_DATA	用于指定数据所附属的地方。例如,图片数据应该附属于联系人
ACTION_EDIT	访问编辑数据
ACTION_PICK	从数据中选择一个子项目,并返回所选中的项目
ACTION_CHOOSER	显示一个 Activity 选择器,允许用户在进程之前选择所要运行的 Activity
ACTION_GET_CONTENT	允许用户选择特殊种类的数据(如,拍一张相片或录一段音),并返回
ACTION_DIAL	拨打一个指定的号码,显示已拨号码的拨号用户界面,等待用户拨出号码
ACTION_CALL	根据指定的数据执行一次呼叫
ACTION_SEND	传递数据,被传送的数据没有指定,接收的 Action 请求用户发数据
ACTION_SENDTO	发送一个信息到指定的某人
ACTION_ANSWER	处理一个打进电话呼叫
ACTION_INSERT	插入一条空项目到已给的容器
ACTION_DELETE	从容器中删除已给的数据
ACTION_RUN	运行数据
ACTION_SYNC	同步执行一个数据
ACTION_PICK_ACTIVITY	为已知的 Intent 选择一个 Activity,返回选中的类
ACTION_SEARCH	执行一次搜索
ACTION_WEB_SEARCH	执行一次 Web 搜索
ACTION_FACTORY_TEST	工场测试的主要进入点

表 7-2　常见的标准的 Activity 广播 Action

标准的广播 Actions 名称	功能与作用
ACTION_TIME_TICK	当前时间改变，每分钟都发送，不能通过组件声明来接收，只有通过 Context. registerReceiver()方法来注册
ACTION_TIME_CHANGED	时间被设置
ACTION_TIMEZONE_CHANGED	时间区改变
ACTION_BOOT_COMPLETED	系统完成启动后，一次广播
ACTION_PACKAGE_ADDED	一个新应用包已经安装在设备上，数据包括包名（最新安装的包程序不能接收到这个广播）
ACTION_PACKAGE_CHANGED	一个已存在的应用程序包已经改变，包括包名
ACTION_PACKAGE_REMOVED	一个已存在的应用程序包已经从设备上移除，包括包名（正在被安装的包程序不能接收到这个广播）
ACTION_PACKAGE_RESTARTED	用户重新开始一个包，包的所有进程将被杀死，所有与其联系的运行时间状态应该被移除，包括包名（重新开始包程序不能接收到这个广播）
ACTION_PACKAGE_DATA_CLEARED	用户已经清除一个包的数据，包括包名（清除包程序不能接收到这个广播）
ACTION_BATTERY_CHANGED	电池的充电状态、电荷级别改变，不能通过组建声明接收这个广播，只有通过 Context. registerReceiver()注册

（2）数据（Data）。Data 为具体的数据，一般由一个 URI 变量来表示。URI 有四个属性 scheme、host、port、path。例如：

```
content://com.shnu.edu.cn:8080/files
                _scheme 部分: content
                _host 部分: com.shnu.edu.cn
                _port 部分: 8080
                _path 部分: files
```

（3）类别（Category）。Category 指定 Action 范围，这个选项指定了将要执行的这个 Action 的其他一些额外的约束。如，所有应用的主 Activity，都需要一个 Category 为 CATEGORY_LAUNCHER、Action 为 ACTION_MAIN 的 Intent。

（4）数据类型（Type）。Type 显式指定 Intent 的数据类型（MIME）。例如 ACTION_VIEW 同时指定为 Type 为 Image，则调出浏览图片的应用。一般 Intent 的数据类型能够根据数据本身进行判定，但是通过设置这个属性，可以强制采用显式指定的类型而不再进行判定。

（5）组件（Component）。Component 指定 Intent 的目标组件的类名称。通常在 Implicit Intents 模式下，Android 会根据 Intent 中包含的其他属性的信息，例如 Action、Data/Type、Category 进行查找，最终找到一个与之匹配的目标组件。但是，如果指定了 Component 这个属性，将直接使用它指定的组件，这种模式叫作 Explicit Intents，适合在明确知道这就是一个内部模块的时候使用它。指定了这个属性以后，Intent 的其他所有属性都是可选的。Component 就是完整的类名，如 com. shnu. MyActivity. class，一旦指明可以直接调用，将不再执行上述查找过程，自然速度快。

(6) 附加信息(Extra)。Extra 是其他所有附加信息的集合。使用 Extra 可以为组件提供扩展信息,例如,如果要执行“发送电子邮件”这个动作,可以将电子邮件的标题、正文等保存在 Extra 里,传给电子邮件发送组件。

(7) 标志信息(Flag)。顾名思义,Flags 是一个整型数,有一些列的标志位构成,这些标志,是用来指明运行模式的。例如,你期望这个意图的执行者与由你运行在两个完全不同的任务中,就需要设置 FLAG_ACTIVITY_NEW_TASK 的标志位。

在 android.content.Indent 中一共定义了 20 种不同的 Flag(标志信息),其中和 Task 紧密关联的有四种:

- FLAG_ACTIVITY_NEW_TASK
- FLAG_ACTIVITY_CLEAR_TOP
- FLAG_ACTIVITY_RESET_TASK_IF_NEEDED
- FLAG_ACTIVITY_SINGLE_TOP

在使用这四个 Flag 时,一个 Intent 可以设置一个 Flag,也可以选择若干个进行组合。

默认情况下,通过 startActivity()启动一个新的 Activity,新的 Activity 将会和调用者在同一个 stack 中。

- 如果在传递给 startActivity()的 Intent 对象里包含了 FLAG_ACTION_NEW_TASK,情况将发生变化,系统将为新的 Activity“寻找”一个不同于调用者的 Task。不过要找的 Task 是不是一定就是 NEW 呢?如果是第一次执行,则这个设想成立,如果不是,也就是说已经有一个包含此 Activity 的 Task 存在,则不会再启动 Activity。
- 如果 Flag 是 FLAG_ACTIVITY_CLEAR_TOP,同时当前的 Task 里已经有了这个 Activity,那么情形又将不一样。Android 不但不会启动新的 Activity 实例,而且还会将 Task 里该 Activity 之上的所有 Activity 一律结束掉,然后将 Intent 发给这个已存在的 Activity。Activity 收到 Intent 之后,可以在 onNewIntent()里做下一步的处理,也可以自行结束然后重新创建自己。
- 如果 Activity 在 AndroidManifest.xml 里将启动模式设置成 multiple 默认模式,并且 Intent 里也没有设置 FLAG_ACTIVITY_SIGNEL_TOP,那么它将选择后者。否则,它将选择前者。FLAG_ACTIVITY_CLEAR_TOP 还可以和 FLAG_ACTION_NEW_TASK 配合使用。
- 如果 Flag 设置的是 FLAG_ACTIVITY_SINGLE_TOP,则意味着如果 Activity 已经是运行的,则该 Activity 将不会再被启动。

下面是一些简单的例子:

```
ACTION_VIEW   content://contacts/1        //显示 identifier 为 1 的联系人的信息
ACTION_DIAL   content://contacts/1        //给这个联系人打电话
ACTION_GET_CONTENT with MIME type vnd.android.cursor.item/phone  //用来列出列表中的所有人
//的电话号码
```

综上所述,action、data/type、category 和 extras 组合在一起可以表达一种确定的语义,如“给某某发送短信”之类语义等。

7.3 Intent 的解析

在应用中，可通过两种形式来使用 Intent。

- 直接 Intent：指定了 component 属性的 Intent（调用 setComponent(ComponentName)或者 setClass(Context, Class)来指定）。通过指定具体的组件类，通知应用启动对应的组件。
- 间接 Intent：没有指定 component 属性的 Intent。这些 Intent 需要包含足够的信息，这样系统才能根据这些信息在所有的可用组件中确定满足此 Intent 的组件。

对于直接 Intent，Android 不需要去做解析，因为目标组件已经很明确，Android 需要解析的是那些间接 Intent，通过解析，将 Intent 映射给可以处理此 Intent 的 Activity、BroadcastReceiver 或 Service。

Intent 解析机制主要是通过查找已注册在 AndroidManifest.xml 中的所有 IntentFilter 及其中定义的 Intent，最终找到匹配的 Intent 的组件。在这个解析过程中，Android 是通过 Intent 的 action、type、category 这三个属性来进行判断的，判断方法如下：

(1) 如果 Intent 指明定了 action，则目标组件的 IntentFilter 的 action 列表中就必须包含有这个 action，否则不能匹配；如果 Intent 没有提供 type，系统将从 data 中得到数据类型。和 action 一样，目标组件的数据类型列表中必须包含 Intent 的数据类型，否则不能匹配。

(2) 如果 Intent 中的数据不是 content:类型的 URI，而且 Intent 也没有明确指定它的 type，将根据 Intent 中数据的 scheme（比如 http:或者 mailto:）进行匹配。同上，Intent 的 scheme 必须出现在目标组件的 scheme 列表中。

(3) 如果 Intent 指定了一个或多个 category，这些类别必须全部出现在组建的类别列表中。例如 Intent 中包含了两个类别：LAUNCHER_CATEGORY 和 ALTERNATIVE_CATEGORY，解析得到的目标组件必须至少包含这两个类别。

7.3.1 动作(Action)样例

Action 在 intent-filter 中设定样例如下。

<intent-filter>元素中可以包括子元素<action>，比如：

```
…
< intent - filter >
      < action android:name = "com.shnu.SHOW_CURRENT" />
      < action android:name = "com.shnu.SHOW_RECENT" />
      < action android:name = "com.shnu.SHOW_PENDING" />
</intent - filter >
…
```

- 一个<intent-filter>元素至少应该包含一个<action>，否则任何 Intent 请求都不能和该<intent-filter>匹配。

- 如果 Intent 请求的 Action 和<intent-filter>中的某一个<action>匹配，那么该 Intent 就通过了这个<intent-filter>的动作测试。
- 如果 Intent 请求或<intent-filter>中没有说明具体的 Action 类型，那么会出现下面两种情况。

(1) 如果<intent-filter>中没有包含任何 Action 类型，那么无论什么 Intent 请求都无法和这个<intent- filter>匹配；

(2) 反之，如果 Intent 请求中没有设定 Action 类型，那么只要<intent-filter>中包含有 Action 类型，这个 Intent 请求就将顺利地通过<intent-filter>的行为测试。

7.3.2 类别(category)样例

AndroidManifest. xml 文件中设定：

<intent-filter>元素可以包含<category>子元素，比如：

```
        < intent - filter . . . >
        < category android:name = "android. Intent. Category. DEFAULT" />
        < category android:name = "android. Intent. Category. BROWSABLE" />
</intent - filter >
```

(1) 只有当 Intent 请求中所有的 Category 与组件中某一个 IntentFilter 的<category>完全匹配时，才会让该 Intent 请求通过测试。IntentFilter 中多余的<category>声明并不会导致匹配失败。

(2) 一个没有指定任何类别测试的 IntentFilter 仅仅只会匹配没有设置类别的 Intent 请求。

7.3.3 数据(data)样例

数据在<intent-filter>中的描述如下：

```
< intent - filter . . . >
        < data android:type = "video/mpeg" android:scheme = "http" . . . />
        < data android:type = "audio/mpeg" android:scheme = "http" . . . />
</intent - filter >
```

元素指定了希望接受的 Intent 请求的数据 URI 和数据类型，URI 被分成三部分来进行匹配：scheme、authority 和 path。其中，用 setData()设定的 Intent 请求的 URI 数据类型和 scheme 必须与 IntentFilter 中所指定的一致。若 IntentFilter 中还指定了 authority 或 path，它们也需要相匹配才会通过测试。

7.4 Intent 的使用

7.4.1 Intent 的构造函数

Intent 有六种构造函数，如表 7-3 所示。

表 7-3 Intent 的六种构造函数

构造函数	说明
Intent()	空构造函数
Intent(Intent o)	复制构造函数
Intent(String action)	指定 Action 类型的构造函数
Intent(String action, Uri uri)	指定 Action 类型和 Uri 的构造函数，URI 主要是结合程序之间的数据共享 ContentProvider
Intent(Context packageContext, Class<?> cls)	传入组件的构造函数，也就是上文提到的
Intent(String action, Uri uri, Context packageContext, Class<?> cls)	前两种结合体

其中，Intent(String action, Uri uri)比较常用，action 就是对应在 AndroidMainfest.xml 中的 action 节点的 name 属性值。如：

使用 Intent 激活 Android 自带的电话拨号程序，创建 Intent 的代码如下所示：

```
Intent i = new Intent(Intent.ACTION_DIAL,Uri.parse("tel://13800138000"));
```

执行"startActivity(i);"代码的时候，系统就会自动打开系统拨号程序(见图 7-3)。

图 7-3 Intent 拨号

7.4.2 常见的 Intent 用例

1. 利用 Intent 在 Activity 之间传递数据

在 MainActivity 中执行如下代码：

```
Bundle bundle = new Bundle();
bundle.putStringArray("name", "Shanghai Normal University");
Intent intent = new Intent(MainActivity.this, ShowActivity.class);
intent.putExtras(bundle);
startActivity(intent);
```

在 ShowActivity 类中，代码如下：

```
Bundle bundle = this.getIntent().getExtras();
String[] arrName = bundle.getStringArray("name");
```

以上代码就实现了 Activity 之间的数据传递!

2. 调出拨打电话界面

```
Uri telUri = Uri.parse("tel:100861");
Intent intent = new Intent(Intent.ACTION_DIAL, telUri);
startActivity(intent);
```

3. 直接拨打电话

```
Uri callUri = Uri.parse("tel:100861");
Intent intent = new Intent(Intent.ACTION_CALL, callUri);
startActivity(intent);
```

注:必须在配置文件中加入<uses-permission id="Android. permission. CALL_PHONE" />。

4. 发短信

```
Uri smsUri = Uri.parse("tel:100861");
Intent intent = new Intent(Intent.ACTION_VIEW, smsUri);
intent.putExtra("sms_body", "this is a test for SMS!!");
intent.setType("vnd.androiddir/mms - sms");
startActivity(intent);
```

5. 发彩信

```
Uri mmsUri = Uri.parse("content://media/external/images/media/23");
Intent intent = new Intent(Intent.ACTION_SEND);
intent.putExtra("sms_body", "MMS Demo");
intent.putExtra(Intent.EXTRA_STREAM, mmsUri);
intent.setType("image/png");
startActivity(intent);
```

6. 调用系统发邮件

```
Uri emailUri = Uri.parse("mailto:liluqun@gmail.com");
Intent intent = new Intent(Intent.ACTION_SENDTO, emailUri);
startActivity(intent);
```

7. 直接发送邮件

```
Intent intent = new Intent(Intent.ACTION_SEND);
String[] tos = { "liluqun@gmail.com" };
String[] ccs = { "liluqun@gmail.com" };
intent.putExtra(Intent.EXTRA_EMAIL, tos);
intent.putExtra(Intent.EXTRA_CC, ccs);
intent.putExtra(Intent.EXTRA_TEXT, "body");
intent.putExtra(Intent.EXTRA_SUBJECT, "subject");
intent.setType("message/rfc882");
Intent.createChooser(returnIt, "Choose Email Client");
startActivity(intent);
```

8. 调用 Web 浏览器

```
Uri myUri = Uri.parse("http://www.shnu.edu.cn");
Intent intent = new Intent(Intent.ACTION_VIEW, myUri);
startActivity(intent);
```

9. 显示地图

```
Uri mapUri = Uri.parse("geo:33.77, -78.55");
Intent intent = new Intent(Intent.ACTION_VIEW, mapUri);
startActivity(intent);
```

10. 播放 mp3 音乐

```
Uri uri = Uri.parse("file:///sdcard/song.mp3");
Intent it = new Intent(Intent.ACTION_VIEW, uri);
it.setType("audio/mp3");
startActivity(it);     //播放多媒体
```

11. 调用系统相机应用程序,并存储拍下来的照片

```
Intent intent = new Intent(MediaStore.ACTION_IMAGE_CAPTURE);
time = Calendar.getInstance().getTimeInMillis();
intent.putExtra(MediaStore.EXTRA_OUTPUT,
 Uri.fromFile(new File(Environment.getExternalStorageDirectory().getAbsolutePath() + "/
tucue", time + ".jpg")));
startActivityForResult(intent, ACTIVITY_GET_CAMERA_IMAGE);
```

12. 卸载 App 软件

```
Uri uninstallUri = Uri.fromParts("package", "xxx", null);
Intent intent = new Intent(Intent.ACTION_DELETE, uninstallUri);
startActivity(intent);
```

13. 安装 App 软件

```
Uri installUri = Uri.fromParts("package", "xxx", null);
Intent intent = new Intent(Intent.ACTION_PACKAGE_ADDED, installUri);
startActivity(intent);
```

7.5 组件通过 Intent 通信方式

其通信方式可以归纳为两种方式:

(1) 点对点通信方式。即:同一个 Android 应用程序中或不同应用程序中的组件,发送 Intent 给其他组件进行通信,这种通信方式可以是单向的或双向的。任意时间参与通信的组件都是一对一的关系。

(2) 广播通信方式。即:Android 应用程序中的组件可以发送广播信息,把 Intent 广播出去;其他 Android 应用程序或组件可以监听该广播消息,并响应完成相应的程序。通信方式是单向的,发送与接收消息的组件是一对一或一对多的关系。

这两种通信方式均可以实现不同应用间组件功能的系统无缝集成(见表 7-4)。

表 7-4　组件通信相关方法

组 件 名 称	方 法 名 称
Activity	startActvity( ) startActvityForResult( )
Service	startService( ) bindService( )
Broadcasts	sendBroadcasts( ) sendOrderedBroadcasts( ) sendStickyBroadcasts( )

以下主要介绍 Activity 和 Broadcasts 通信方式,Service 将会单独在第 8 章中介绍。

7.6　组件的点对点通信方式

Android 系统中,大多数应用程序都包含一个或多个 Activity,同一应用或不同应用程序的 Activity 可以通过 Intent 进行切换和数据传输。任意时间参与通信的组件都是一对一的关系。按照 Intent 启动 Activity 的方式可以分为显示启动和隐式启动两种方式。

7.6.1　显式启动 Activity

使用 Intent 显式启动方法,只需在一个 Activity 中创建一个 Intent 对象,并在这个 Activity 中调用 startActivity(Intent)即可!这种方法虽然简单,但是两个 Activity 之间的调用关系在程序代码中已经固定,Activity 之间耦合度高。

```
Intent intent = new Intent(Activity_A.this,Activity_B.class);
startActivity(intent);
```

【例 7-1】 教学案例设计:两个 Activity 之间页面切换,实现一个 Activity 的数据传递给另一个 Activity。在一个应用程序中分别设计两个 Activity:Activity_A 和 Activity_B,Activity_A 中使用 Intent,启动 Activity_B;并在 Activity_B 上显示 Activity_A 传递过来的字符串。

Activity_A.java:

```
package com.shnu.explicitforstartactivity;
import android.os.Bundle;
import android.view.View;
import android.app.Activity;
import android.content.Intent;

public class Activity_A extends Activity {
    public static String INTENT_DATA = "intent_data";
```

```
    @Override
    protected void onCreate(Bundle savedInstanceState) {
        super.onCreate(savedInstanceState);
        setContentView(R.layout.activity_a);
        findViewById(R.id.activity_a_btn).setOnClickListener(new View.OnClickListener()
    {
            @Override
            public void onClick(View v) {
                // TODO Auto-generated method stub
                Intent intent = new Intent(Activity_A.this, Activity_B.class);
                intent.putExtra(INTENT_DATA, "from Activity_A");
                startActivity(intent);
            }
        });
    }
```

Activity_a.xml:

```
<RelativeLayout xmlns:android="http://schemas.android.com/apk/res/android"
    xmlns:tools="http://schemas.android.com/tools"
    android:layout_width="match_parent"
    android:layout_height="match_parent"
    tools:context=".MainActivity" >

    <Button
        android:id="@+id/activity_a_btn"
        android:layout_width="wrap_content"
        android:layout_height="wrap_content"
        android:layout_centerHorizontal="true"
        android:layout_centerVertical="true"
        android:text="@string/hello_world" />

</RelativeLayout>
```

Activity B.java:

```
package com.shnu.explicitforstartactivity;

import android.os.Bundle;
import android.widget.TextView;
import android.app.Activity;
import android.content.Intent;

public class Activity_B extends Activity {
    private TextView activity_b_txt;
    @Override
    protected void onCreate(Bundle savedInstanceState) {
        super.onCreate(savedInstanceState);
        setContentView(R.layout.activity_b);
        activity_b_txt = (TextView)findViewById(R.id.activity_b_txt);
        Intent intent = getIntent();
```

```
        String fromIntent =  intent.getStringExtra(Activity_A.INTENT_DATA);
        activity_b_txt.setText(fromIntent);
    }

}
```

Activity_b.xml：

```
<RelativeLayout xmlns:android="http://schemas.android.com/apk/res/android"
    xmlns:tools="http://schemas.android.com/tools"
    android:layout_width="match_parent"
    android:layout_height="match_parent"
    tools:context=".MainActivity" >

    <TextView
        android:id="@+id/activity_b_txt"
        android:layout_width="wrap_content"
        android:layout_height="wrap_content"
        android:layout_centerHorizontal="true"
        android:layout_centerVertical="true"
        android:text="@string/hello_world" />

</RelativeLayout>
```

AndroidManifest.xml：

```
<?xml version="1.0" encoding="utf-8"?>
<manifest xmlns:android="http://schemas.android.com/apk/res/android"
    package="com.shnu.explicitforstartactivity"
    android:versionCode="1"
    android:versionName="1.0" >

    <uses-sdk
        android:minSdkVersion="8"
        android:targetSdkVersion="17" />

    <application
        android:allowBackup="true"
        android:icon="@drawable/ic_launcher"
        android:label="@string/app_name"
        android:theme="@style/AppTheme" >
        <activity
            android:name="com.shnu.explicitforstartactivity.Activity_A"
            android:label="@string/app_name" >
            <intent-filter>
                <action android:name="android.intent.action.MAIN" />

                <category android:name="android.intent.category.LAUNCHER" />
            </intent-filter>
        </activity>
```

```
        <activity
            android:name=".Activity_B"/>

    </application>

</manifest>
```

运行结果见图 7-4。

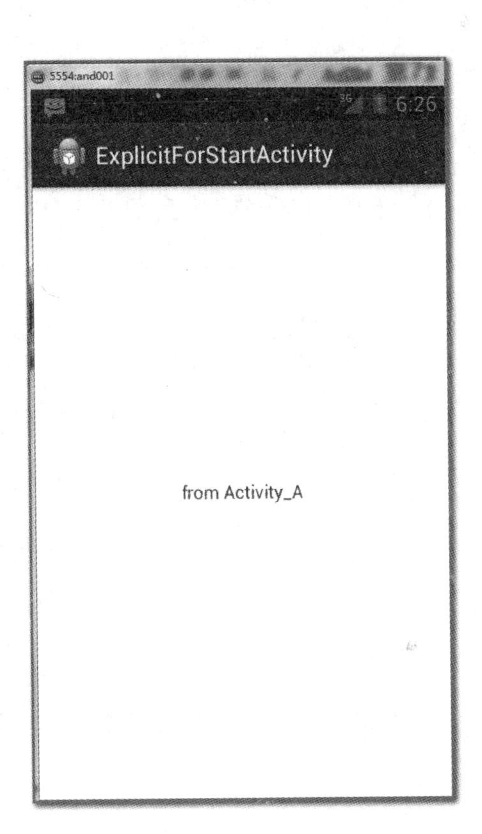

图 7-4　Activity_A 调用 Activity_B 并传送字符串

Activity_a 运行，出现界面 Hello World！按钮，单击该按钮 Activity_b 启动，然后获取从 Activity_a 传过来的字符串“from Activity_A”并显示。

7.6.2　隐式启动 Activity

在一个 Activity 中使用 Intent 隐式启动另一个 Activity 的方法是：在一个 Activity 中创建一个 intent 对象，该对象只指明要完成的 Action、Url、数据即可，其他工作由 Android 操作系统来检索与 Intent 匹配的 Activity（可能有多个 Activity 符合条件），启动符合条件的 Activity。这种方法两个 Activity 之间耦合度比较低。

教学案例设计：演示一个 Activity 隐式调用另一个 Activity。

在一个应用程序中分别设计两个 Activity：Activity_A 和 Activity_B，Activity_A 中使用 Intent，隐式启动 Activity_B；并在 Activity_B 上显示 Activity_A 传递过来的字符串。

【例 7-2】 Activity_A. java。

```
package com. shnu. implicitforstatrtactivity;

import android. os. Bundle;
import android. view. View;
import android. app. Activity;
import android. content. Intent;

public class Activity_A extends Activity {
    public static String MY_ACTION = "android、intent、my_action"
    @Override
    protected void onCreate(Bundle savedInstanceState) {
        super. onCreate(savedInstanceState);
        setContentView(R. layout. activity_a);
        findViewById(R. id. activity_a_btn). setOnClickListener(new View. OnClickListener()
    {
            @Override
            public void onClick(View v) {
                // TODO Auto - generated method stub
                Intent intent = new Intent();
                intent. setAction(MY_ACTION);
                startActivity(intent);
            }
        });
    }
```

Activity_a. xml：

```
<RelativeLayout xmlns:android = "http://schemas. android. com/apk/res/android"
    xmlns:tools = "http://schemas. android. com/tools"
    android:layout_width = "match_parent"
    android:layout_height = "match_parent"
    tools:context = ". MainActivity" >

    <Button
        android:id = "@ + id/activity_a_btn"
        android:layout_width = "wrap_content"
        android:layout_height = "wrap_content"
        android:layout_centerHorizontal = "true"
        android:layout_centerVertical = "true"
        android:text = "@string/hello_world" />

</RelativeLayout>
```

Activity_b. java：

```
package com. shnu. implicitforstatrtactivity;

import android. os. Bundle;
```

```
import android.widget.TextView;
import android.app.Activity;
import android.content.Intent;

public class Activity_B extends Activity {
    private TextView activity_b_txt;
    @Override
    protected void onCreate(Bundle savedInstanceState) {
        super.onCreate(savedInstanceState);
        setContentView(R.layout.activity_b);
        activity_b_txt = (TextView)findViewById(R.id.activity_b_txt);
        Intent intent = getIntent();
        String action = intent.getAction();
        activity_b_txt.setText(action);
    }

}
```

Activity_b.xml:

```
<RelativeLayout xmlns:android = "http://schemas.android.com/apk/res/android"
    xmlns:tools = "http://schemas.android.com/tools"
    android:layout_width = "match_parent"
    android:layout_height = "match_parent"
    tools:context = ".MainActivity" >

    <TextView
        android:id = "@ + id/activity_b_txt"
        android:layout_width = "wrap_content"
        android:layout_height = "wrap_content"
        android:layout_centerHorizontal = "true"
        android:layout_centerVertical = "true"
        android:text = "@string/hello_world" />

</RelativeLayout >
```

AndroidManifest.xml:

```
<?xml version = "1.0" encoding = "utf - 8"?>
<manifest xmlns:android = "http://schemas.android.com/apk/res/android"
    package = "com.shnu.implicitforstatrtactivity"
    android:versionCode = "1"
    android:versionName = "1.0" >

    <uses - sdk
        android:minSdkVersion = "8"
        android:targetSdkVersion = "17" />

    <application
        android:allowBackup = "true"
```

```
        android:icon = "@drawable/ic_launcher"
        android:label = "@string/app_name"
        android:theme = "@style/AppTheme" >
        <activity
            android:name = "com.shnu.implicitforstatrtactivity.Activity_A"
            android:label = "@string/app_name" >
            <intent - filter >
                <action android:name = "android.intent.action.MAIN" />

                <category android:name = "android.intent.category.LAUNCHER" />
            </intent - filter >
        </activity >
         <activity
            android:name = "com.shnu.implicitforstatrtactivity.Activity_B"
             >
            <intent - filter >
                <action android:name = "android.intent.my_action" />
                <category android:name = "android.intent.category.DEFAULT" />
            </intent - filter >
        </activity >
    </application >

</manifest >
```

程序运行结果见图 7-5。

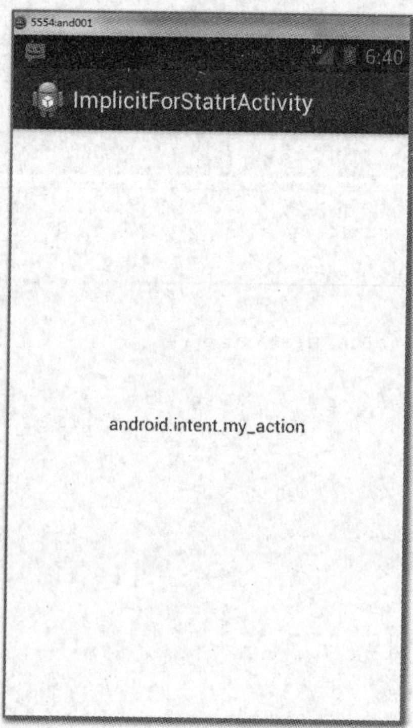

图 7-5　Activity_A 隐式启动 Activity_B

AndroidManifest. xml 文件中 Activity_b 的 Intent-filter 项注册了：

```
<action android:name = "android. intent. my_action" />
```

Activity_a 在程序中通过 Intent 设定 action 为"android. intent. my_action"，可以隐式启动 Activity_b。

7.6.3 强制用户选择启动 Activity

【例 7-3】 教学案例设计：演示一个 Activity 强制用户选择目标 Activity。

设计一个 Activity，利用 Intent 将这个 Activity 中的文字、图片共享给 QQ 或微信好友。

MainActivity. java。

```
package com. shnu. enforceforstartactivity;

import android. app. Activity;
import android. content. Intent;
import android. os. Bundle;
import android. view. View;

public class MainActivity extends Activity {

    @Override
    protected void onCreate(Bundle savedInstanceState) {
        super. onCreate(savedInstanceState);
        setContentView(R. layout. activity_main);
        findViewById(R. id. activity_btn). setOnClickListener(new View. OnClickListener() {

            @Override
            public void onClick(View v) {
                // TODO Auto - generated method stub
             Intent intent = new Intent(Intent. ACTION_SEND);
             intent. setType("text/plain");
             intent. putExtra(Intent. EXTRA_SUBJECT, "分享");
             intent. putExtra(Intent. EXTRA_TEXT, "你好 ");
             intent. putExtra(Intent. EXTRA_TITLE, "我是标题");
             intent. setFlags(Intent. FLAG_ACTIVITY_NEW_TASK);
             startActivity(Intent. createChooser(intent, "请选择"));
            }
        });
    }
}
```

Avtivity_main. xml：

```
<RelativeLayout xmlns:android = "http://schemas. android. com/apk/res/android"
    xmlns:tools = "http://schemas. android. com/tools"
    android:layout_width = "match_parent"
```

```
        android:layout_height = "match_parent"
        tools:context = ".MainActivity" >

        <Button
            android:id = "@ + id/activity_btn"
            android:layout_width = "wrap_content"
            android:layout_height = "wrap_content"
            android:layout_centerHorizontal = "true"
            android:layout_centerVertical = "true"
            android:text = "@string/hello_world" />

</RelativeLayout>
```

AndroidManifest.xml：

```
<?xml version = "1.0" encoding = "utf - 8"?>
<manifest xmlns:android = "http://schemas.android.com/apk/res/android"
    package = "com.shnu.enforceforstartactivity"
    android:versionCode = "1"
    android:versionName = "1.0" >

    <uses - sdk
        android:minSdkVersion = "8"
        android:targetSdkVersion = "17" />

    <application
        android:allowBackup = "true"
        android:icon = "@drawable/ic_launcher"
        android:label = "@string/app_name"
        android:theme = "@style/AppTheme" >
        <activity
            android:name = "com.shnu.enforceforstartactivity.MainActivity"
            android:label = "@string/app_name" >
            <intent - filter >
                <action android:name = "android.intent.action.MAIN" />

                <category android:name = "android.intent.category.LAUNCHER" />
            </intent - filter >
        </activity >
    </application >

</manifest >
```

当用户运行 MainActivity，单击“hello world!”按钮，在系统出现的对话框中选择相应的 Activity，选择微信，就可以把用户在 MainActivity 中的字符串“你好”传递到微信用户的好友对话界面中（见图 7-6）。

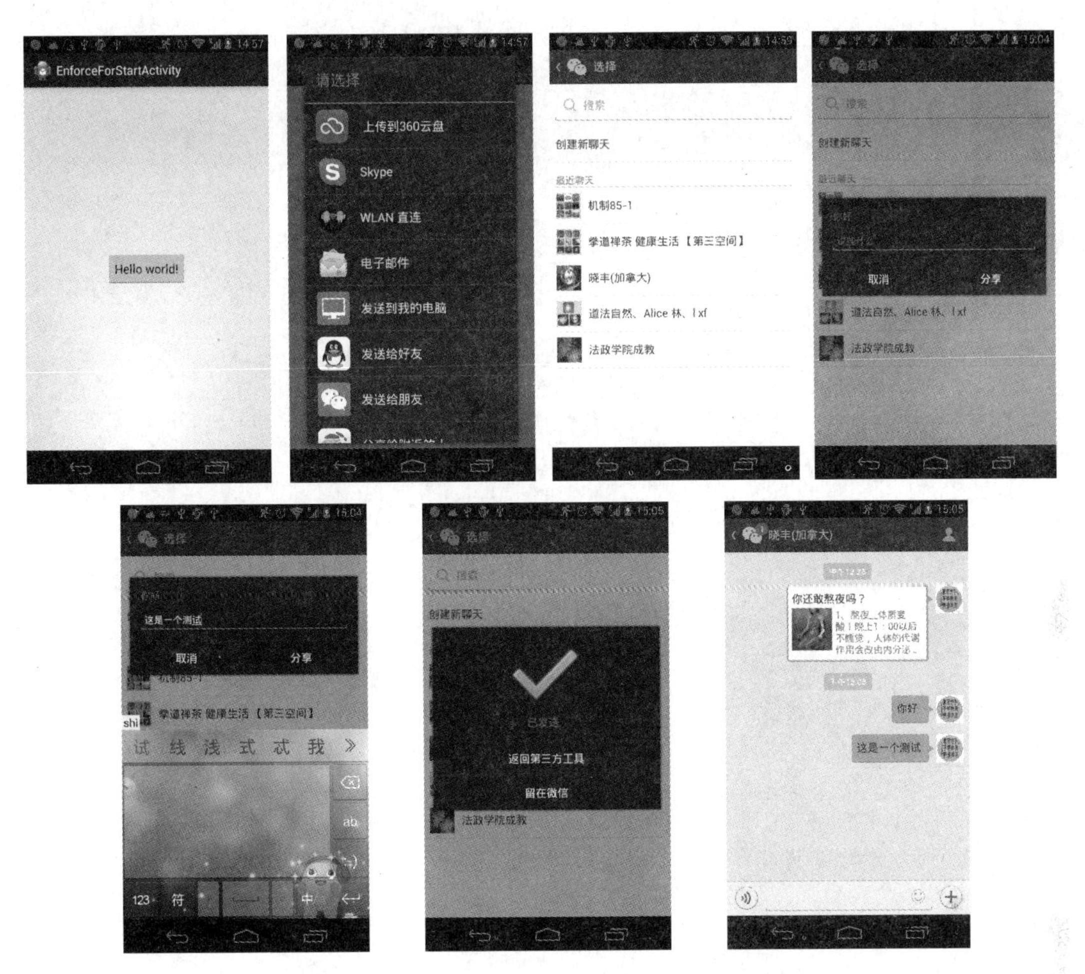

图 7-6　例 7-3 运行效果

7.6.4　获取启动 Activity 的返回值

【例 7-4】　教学案例设计：两个 Activity 之间页面切换，实现两个 Activity 之间的数据传递。

在一个应用程序中分别设计两个 Activity：Activity_A 和 Activity_B，Activity_A 中使用 Intent，启动 Activity_B。并在 Activity_B 上显示 Activity_A 传递过来的字符串，Activity_B 处理字符串后结束，并把结果返回 Activity_A。

Activity_a.java。

```
package com.shnu.startactivityforresult;

import android.app.Activity;
import android.content.Intent;
import android.os.Bundle;
import android.view.View;
```

```
import android.widget.TextView;

public class Activity_A extends Activity {
    public static String INTENT_DATA = "intent_data";
    public static int RESULTCODE = 5;
    private TextView activity_a_txt;
    @Override
    protected void onCreate(Bundle savedInstanceState) {
        super.onCreate(savedInstanceState);
        setContentView(R.layout.activity_a);
        activity_a_txt = (TextView)findViewById(R.id.activity_a_txt);
        findViewById(R.id.activity_a_btn).setOnClickListener(new View.OnClickListener()
    {
            @Override
            public void onClick(View v) {
                // TODO Auto-generated method stub
                Intent intent = new Intent(Activity_A.this,Activity_B.class);
                //第二个参数为 requestCode
                startActivityForResult(intent, 0);
            }
        });
    }

    @Override
    protected void onActivityResult(int requestCode, int resultCode, Intent data) {
        // TODO Auto-generated method stub
        super.onActivityResult(requestCode, resultCode, data);
        if(resultCode == RESULTCODE){
            String result = data.getExtras().getString(INTENT_DATA);
            activity_a_txt.setText(result);
        }

    }
```

Activity_a.xml：

```
<RelativeLayout xmlns:android="http://schemas.android.com/apk/res/android"
    xmlns:tools="http://schemas.android.com/tools"
    android:layout_width="match_parent"
    android:layout_height="match_parent"
    tools:context=".MainActivity" >

    <Button
        android:id="@+id/activity_a_btn"
        android:layout_width="wrap_content"
        android:layout_height="wrap_content"
        android:layout_centerHorizontal="true"
        android:layout_centerVertical="true"
```

```
        android:text = "@string/hello_world" />
    <TextView
        android:id = "@ + id/activity_a_txt"
        android:layout_width = "wrap_content"
        android:layout_height = "wrap_content"
        android:layout_below = "@id/activity_a_btn"
        android:layout_marginTop = "20dip"
        android:layout_centerHorizontal = "true"
        android:text = "@string/hello_world" />

</RelativeLayout>
```

Activity_b.java：

```
package com.shnu.startactivityforresult;

import android.os.Bundle;
import android.view.View;
import android.app.Activity;
import android.content.Intent;

public class Activity_B extends Activity {
    @Override
    protected void onCreate(Bundle savedInstanceState) {
        super.onCreate(savedInstanceState);
        setContentView(R.layout.activity_b);
        findViewById(R.id.activity_b_btn).setOnClickListener(new View.OnClickListener()
    {
                @Override
                public void onClick(View v) {
                    // TODO Auto - generated method stub
                Intent intent = new Intent();
                intent.putExtra(Activity_A.INTENT_DATA, "this is from Activity_B");
                //第一个参数为 resultCode
                setResult(Activity_A.RESULTCODE, intent);
                finish();
                }
            });;
    }

}
```

Activity_b.xml：

```
<RelativeLayout xmlns:android = "http://schemas.android.com/apk/res/android"
    xmlns:tools = "http://schemas.android.com/tools"
    android:layout_width = "match_parent"
    android:layout_height = "match_parent"
    tools:context = ".MainActivity" >
```

```
    < Button
        android:id = "@ + id/activity_b_btn"
        android:layout_width = "wrap_content"
        android:layout_height = "wrap_content"
        android:layout_centerHorizontal = "true"
        android:layout_centerVertical = "true"
        android:text = "返回 Activity_A" />

</RelativeLayout >
```

AndroidManifest.xml：

```
<?xml version = "1.0" encoding = "utf - 8"?>
< manifest xmlns:android = "http://schemas.android.com/apk/res/android"
    package = "com.shnu.startactivityforresult"
    android:versionCode = "1"
    android:versionName = "1.0" >

    < uses - sdk
        android:minSdkVersion = "8"
        android:targetSdkVersion = "17" />

    < application
        android:allowBackup = "true"
        android:icon = "@drawable/ic_launcher"
        android:label = "@string/app_name"
        android:theme = "@style/AppTheme" >
        < activity
            android:name = "com.shnu.startactivityforresult.Activity_A"
            android:label = "@string/app_name" >
            < intent - filter >
                < action android:name = "android.intent.action.MAIN" />

                < category android:name = "android.intent.category.LAUNCHER" />
            </intent - filter >
        </activity >
         < activity
              android:name = "com.shnu.startactivityforresult.Activity_B"/>
    </application >

</manifest >
```

运行结果：

Activity_A 运行，用户单击“Hello world!”按钮，启动 Activity_B；用户单击“返回 Activity_A”按钮，返回到 Activity_A 用户界面并将 Activity_B 中的字符串显示在 Activity_A 界面中（见图 7-7）。

图 7-7　例 7-4 带返回值 Activity 运行效果图

7.7　广播通信——组件的一对多通信方式

Android 系统中，组件之间的通信还可以采用 Intent 发送广播消息方法实现，其他组件监听并接收广播，来实现组件之间的通信。在 Android 里面有各种各样的广播，比如电池的使用状态，电话的接收和短信的接收都会产生一个广播，其他应用组件可以监听这些广播并做出响应处理，应用程序也可以发送广播。

由于发送广播消息的通常由一个应用或组件来完成，而接收广播消息的应用或组件可以是一个或多个，所以可以认为这种组件通信方式是一对多通信方式。这种通信方式的组件之间的应用集成耦合度更低。

7.7.1　自定义广播消息的发送和接收

一个 Android 系统在运行过程中，不同的应用组件可以使用 Intent 发送广播消息(当然必须有相关权限)，Android 系统自带的组件也会发送广播消息，如系统启动完成、系统收到短信、电池电量过低、网络连接发生变化等等。其他 BoardcastReceiver 组件如果注册监听相关广播，则可以收到相应的广播消息，并进行处理。

- 广播消息的发送。广播消息的发送非常简单，只需定义一个消息名称字符串(最好全局唯一，建议可以使用应用程序的包名，不要与系统消息字符串名称冲突)，使用该字符串创建一个 Intent，并使用 Intent 的 putExtra()方法在广播消息中添加数据，然后调用 sendBroadcast()方法将广播消息发送出去。
- 广播消息的接收。通过定义一个继承 BroadcastReceiver 类来实现，继承该类并覆盖其 onReceiver()方法，在该方法中响应事件。另外，需要在 androdManifest. xml 中，注册监听的广播。如：

```
<receiver
        android:name = "com.shnu.boardcastdemo.MyReceiver"
        android:enabled = "true"
        android:exported = "true" >
        <intent - filter >
            <action android:name = "com.shnu.broadcast.action" />

        </intent - filter >
    </receiver >
```

或者在程序代码中注册：

```
IntentFilter filter = new IntentFilter(ACTION_INTENT_TEST);
MyReceiver receiver = new MyReceiver();
registerReceiver(receiver,filter);
```

注册 BroadcastReceiver 的应用程序不需要一直运行，当 Android 系统接收到与之匹配的广播消息时，会自动启动此 BroadcastReceiver，因此 BroadcastReceiver 非常适合做一些资源管理工作。在 BroadcastReceiver 接收到与之匹配的广播消息后，onReceive()方法会被调用，但 onReceive()方法必须要在 5 秒钟执行完毕，否则 Android 系统会认为该组件失去响应，并提示用户强行关闭该组件。

【例 7-5】 教学案例设计：设计两个 Android 应用程序，一个程序发送自定义广播消息"com. shnu. action"，另一个程序接收广播消息"com. shnu. action"，并显示消息内容。

MainActivity. java：

```
package com.shnu.boardcastdemo;

import android.os.Bundle;
import android.app.Activity;
import android.content.Intent;
import android.content.IntentFilter;

import android.view.View;
import android.view.View.OnClickListener;
import android.widget.Button;
import android.widget.Toast;

public class MainActivity extends Activity {

    public static final String ACTION_INTENT_TEST = "com.shnu.broadcast.action";

    /** Called when the activity is first created. */
    @Override
    public void onCreate(Bundle savedInstanceState) {
        super.onCreate(savedInstanceState);
        setContentView(R.layout.activity_main);
        Button btn = (Button) findViewById(R.id.button1);
        btn.setOnClickListener(new OnClickListener() {
```

```
            @Override
            public void onClick(View v) {
                // TODO Auto - generated method stub
                Intent intent = new Intent(ACTION_INTENT_TEST);
                intent.putExtra("content", "Hello World from Shnu");
                sendBroadcast(intent);
                String msg = intent.getAction() + "is broadcasted!!";
                System.out.println(msg);
                // mtent 的动作名称
                // IntentFilter filter = new IntentFilter(ACTION_INTENT_TEST);
                // MyReceiver receiver = new MyReceiver();
                // registerReceiver(receiver,filter);

            }
        });
    }
}
```

BroadcastReceiver.java：

```
package com.shnu.boardcastdemo;

import android.content.BroadcastReceiver;
import android.content.Context;
import android.content.Intent;
import android.widget.Toast;

public class MyReceiver extends BroadcastReceiver {
    public MyReceiver() {
    }

    @Override
    public void onReceive(Context context, Intent intent) {
        // TODO Auto - generated method stub

        String s = intent.getAction() + " is received!!" + "\n Content is: "
                    + intent.getStringExtra("content");
        Toast.makeText(context, s, Toast.LENGTH_LONG).show();

    }
```

Activity_main.xml：

```
<RelativeLayout xmlns:android = "http://schemas.android.com/apk/res/android"
    xmlns:tools = "http://schemas.android.com/tools"
    android:layout_width = "match_parent"
    android:layout_height = "match_parent"
    android:paddingBottom = "@dimen/activity_vertical_margin"
    android:paddingLeft = "@dimen/activity_horizontal_margin"
```

```
    android:paddingRight = "@dimen/activity_horizontal_margin"
    android:paddingTop = "@dimen/activity_vertical_margin"
    tools:context = ".MainActivity" >

    <TextView
        android:id = "@+id/textView1"
        android:layout_width = "wrap_content"
        android:layout_height = "wrap_content"
        android:text = "@string/hello_world" />

    <Button
        android:id = "@+id/button1"
        android:layout_width = "wrap_content"
        android:layout_height = "wrap_content"
        android:layout_below = "@+id/textView1"
        android:layout_marginTop = "88dp"
        android:layout_toRightOf = "@+id/textView1"
        android:text = "Button" />

</RelativeLayout>
```

AndroidManifest.xml：

```
<?xml version = "1.0" encoding = "utf-8"?>
<manifest package = "com.shnu.boardcastdemo"
    android:versionCode = "1"
    android:versionName = "1.0" xmlns:android = "http://schemas.android.com/apk/res/android">

    <uses-sdk
        android:minSdkVersion = "8"
        android:targetSdkVersion = "17" />

    <application
        android:allowBackup = "true"
        android:icon = "@drawable/ic_launcher"
        android:label = "@string/app_name"
        android:theme = "@style/AppTheme" >
        <activity
            android:name = "com.shnu.boardcastdemo.MainActivity"
            android:label = "@string/app_name" >
            <intent-filter>
                <action android:name = "android.intent.action.MAIN" />

                <category android:name = "android.intent.category.LAUNCHER" />
            </intent-filter>
        </activity>

        <receiver
            android:name = "com.shnu.boardcastdemo.MyReceiver"
```

```
            android:enabled = "true"
            android:exported = "true" >
            <intent - filter>
                <action android:name = "com.shnu.broadcast.action" />


            </intent - filter>
        </receiver>
    </application>

</manifest>
```

软件运行效果见图 7-8。

图 7-8　例 7-5 运行效果图

7.7.2　系统广播消息的接收

在 Android 系统中，系统运行的每个状态都有许多广播消息在发送，应用程序可以通过一个派生 BoradcastReceiver 来创建一个广播消息接收类，用来监听广播消息，并在 onReceive()方法处理相关事件，另外要监听的广播消息名称一定要在 AndroidManifest.xml 文件或程序代码中注册。

【例 7-6】 教学案例设计：设计一个 Android 自启动软件，当 Android 手机开机完成后，该软件自动运行，并显示“成功完成自启动”字符串。

MyReceiver. java：

```
package com.example.broadcastreceiver;

import android.content.BroadcastReceiver;
import android.content.Context;
import android.content.Intent;

public class MyReceiver extends BroadcastReceiver {
    static final String ACTION = "android.intent.action.BOOT_COMPLETED";

    @Override
    public void onReceive(Context context, Intent intent) {
        if (intent.getAction().equals(ACTION)) {
                Intent sayHelloIntent = new Intent(context, MainActivity.class);

                context.startActivity(sayHelloIntent);

        }
    }
}
```

MainActivity. java：

```
package com.example.broadcastreceiver;

import android.os.Bundle;
import android.app.Activity;
import android.view.Menu;
import android.widget.TextView;

public class MainActivity extends Activity {

    @Override
    protected void onCreate(Bundle savedInstanceState) {
        super.onCreate(savedInstanceState);
        setContentView(R.layout.activity_main);
        TextView tv = new TextView(this);
         tv.setText("成功完成自启动!");
        setContentView(tv);

    }
}
```

Activity_main. xml：

```
<RelativeLayout xmlns:android = "http://schemas.android.com/apk/res/android"
    xmlns:tools = "http://schemas.android.com/tools"
    android:layout_width = "match_parent"
    android:layout_height = "match_parent"
```

```
    android:paddingBottom = "@dimen/activity_vertical_margin"
    android:paddingLeft = "@dimen/activity_horizontal_margin"
    android:paddingRight = "@dimen/activity_horizontal_margin"
    android:paddingTop = "@dimen/activity_vertical_margin"
    tools:context = ".MainActivity" >

    <TextView
        android:layout_width = "wrap_content"
        android:layout_height = "wrap_content"
        android:text = "@string/hello_world" />

</RelativeLayout>
```

AndroidManifest.xml：

```
<?xml version = "1.0" encoding = "utf-8"?>
<manifest xmlns:android = "http://schemas.android.com/apk/res/android"
    package = "com.example.broadcastreceiver"
    android:versionCode = "1"
    android:versionName = "1.0" >

    <uses-sdk
        android:minSdkVersion = "8"
        android:targetSdkVersion = "17" />

    <application
        android:allowBackup = "true"
        android:icon = "@drawable/ic_launcher"
        android:label = "@string/app_name"
        android:theme = "@style/AppTheme" >
        <activity
            android:name = "com.example.broadcastreceiver.MainActivity"
            android:label = "@string/app_name" >
            <intent-filter>
                <action android:name = "android.intent.action.MAIN" />

                <category android:name = "android.intent.category.LAUNCHER" />
            </intent-filter>
        </activity>

        <receiver
            android:name = "com.example.broadcastreceiver.MyReceiver"
            android:enabled = "true"
            android:exported = "true" >
            <intent-filter>
                <action android:name = "android.intent.action.BOOT_COMPLETED" />


            </intent-filter>
```

```
        </receiver>
    </application>

</manifest>
```

软件运行结果见图 7-9。

图 7-9　例 7-6 运行结果图

7.8　习题与课外阅读

7.8.1　习题

(1) 简述 Intent 的功能。

(2) 编写一个程序利用 Intent 发送短信。

(3) 编写一个程序利用 Intent 打开摄像头拍照。

(4) 简述 Android 广播机制。Broadcast 有哪两种监听注册方式?

(5) 编写一个 BroadcastReceiver 程序,当收到短信时候,使用 Toast 显示已经收到短信。

7.8.2　课外阅读

(1) 了解一下 Android 系统中常见的广播消息名称。

(2) 了解一下如何利用 Intent 做百度或谷歌地图服务路径规划。

第8章 Service与后台服务

Service是Android系统的四大核心组件之一，Service是一种没有界面的组件，同我们平常在Windows或Linux系统中所理解的“服务”一样，Android上的Service也是运行在后台的，主要用来完成一些逻辑功能较复杂、耗时的运算，其运行时间可以从系统启动到系统关闭为止。通过本章的学习可以让读者理解Service的基本原理，掌握Service的使用与创建，理解Android系统进程之间的通信机制。

本章学习目标：

- 了解Service的原理及应用，Service的生命周期；
- 掌握Service的基本创建、启动、停止；
- 如何将已有的逻辑业务代码封装成Service；
- 掌握如何将Service的数据与UI通信；
- 掌握AIDL语言的基本使用方法以及Service的创建。

8.1 Service简介

Service是Android系统中四大核心组件之一，主要用于两个目的：后台运行和跨进程访问。Service可以在不显示界面的前提下在后台运行指定的任务，这样可以不影响用户做其他事情。Service无UI交互界面，运行在系统后台，而前台可以是不同的应用组件在运行。Service允许被其他组件绑定，来完成进程内、跨进程通信，实现应用无缝集成。通常Service用来完成许多耗时和计算量较大的运算，如网络连接、音乐播放、文件输入输出、读写Content provider等。Service的优先级比Activity高，当系统运行资源不足时，Service不会被Android系统优先终止。即使Service被终止，当系统一旦有足够资源的时候，Service将自动恢复运行。Service除了在后台提供服务外，和可以用于进程之间通信(IPC)，实现不同Android应用进程之间通信。

按照Service的启动和使用方式，可以分三种：

(1) 直接启动使用方式，即startService()启动的服务。这种启动方式需要Android系统中的组件(如Activity)通过创建启动Service的Intent，然后调用方法Context.startService()，启动Service。Service启动后，通常没有返回值返回给调用它的组件。即使启动该Service的应用或组件经退出，Service仍然可以继续运行。启动后的Service可以在系统中长时间运行，如需终止Service可以通过调用Context.stopService()或Service.stopSelf()方法来完成。

（2）绑定方式启动使用方式，即 bindService()启动的服务。这种启动方式需要 Android 系统中的组件(如 Activity)调用 Context. bindService()方法。Service 与绑定它的组件采用客户机/服务器(C/S)运算模式，前台组件把请求发送给 Service，Service 处理完将结果返回给前台组件。一个 Service 可以与多个前台组件同时绑定，当没有前台组件与 Service 绑定时，Service 即刻终止(destroyed)。

（3）通过 startService()启动使用，又通过绑定 bindService()启动使用方式。这种启动方式可能 Android 系统中的组件(如 Activity)通过创建启动 Service 的 Intent，然后调用方法 Context. startService()，启动 Service；Android 系统中的其他组件(如 Activity)通过调用 Context. bindService()方法启动 Service，当没有前台组件与 Service 绑定时，Service 不会被终止，如需终止，Service 可以通过调用 Context. stopService()或 Service. stopSelf()方法来完成。

8.2 Service 与 Thread 的区别

既然 Service 可以运行在当前进程或独立进程的主线程上，为什么不直接采用线程来实现 Service 的功能呢？

这主要与 Android 的系统机制有关，虽然在 Android 应用中可以非常方便地创建一个线程 Thread，Thread 与 Android 系统 UI 线程相互独立，如：即使一个 Activity 被调用 finish() 终止之后，Thread 仍然有可能继续运行，这样会导致不再持有该 Thread 的引用，无法再控制该 Thread，另外，不同的 Activity 中对同一 Thread 进行控制也比较困难。

举个例子：如果一个 Thread 需要每隔一段时间就要连接网络后台服务器获取数据，那么该 Thread 需要在 Activity 没有运行 start()方法的时候也可以运行。这样就没有办法在该 Activity 里面控制该 Thread。

在这种情况下，创建并启动一个 Service，在 Service 里面完成创建、运行并控制该 Thread 的业务逻辑，Activity 再与这个 Service 通信便解决了该问题(因为任何 Activity 都可以控制同一 Service，而系统也只会创建一个对应 Service 的实例)。

Service 可以被认为是 Android 系统控制线程的工具或者是一种消息服务，而你可以在任何有 Context 的地方调用 Context. startService()、Context. stopService()、Context. bindService()、Context. unbindService()来控制它，也可以在 Service 里注册 BroadcastReceiver，在其他地方通过发送广播来控制它，当然这些都是 单纯使用 Thread 做不到的。

8.3 Service 的创建

编写 Service 代码比较简单，创建最少代码集合的 Service，只需要继承 Service，实现 Service 中的抽象方法 onBind(Intent intent)，方法的返回值是一个 IBinder 对象，该方法主要建立其他组件与 Service 通信的通道，其他组件可以通过 IBinder 对象来获取 Service 引用，来控制使用 Service 逻辑功能，另外，在 AndroidManifest. xml 文件中还要注册这个 Service。一个 Service 的最小代码如下：

```
import android.app.Service;
import android.content.Intent;
import android.os.IBinder;

public class MyService extends Service{
    public IBinder onBind(Intent intent) {
        return null;
    }
}
```

8.4 Service的生命周期

相对于Activity，Service的生命周期稍微复杂些。虽然Service的生命周期密切相关的回调函数比Activity生命周期相关的函数要少，只有onCreate()、onStart()、onDestroy()、onBind()和onUnbind()五个相关回调函数，但是由于Service的启动方式不同，对Service生命周期的回调函数调用方式和影响是不一样的。

(1) 通过startService()方式启动的Service生命周期。

Service会经历onCreate()到onStart()，然后处于运行状态，stopService()的时候调用onDestroy()方法。如果是调用者自己直接退出而没有调用stopService()，Service会一直在后台运行。

(2) 通过bindService()方式启动的Service生命周期。

Service会运行onCreate()，然后调用onBind()，这个时候调用者和Service绑定在一起。若调用者退出，Srevice就会调用onUnbind()－>onDestroyed()方法。调用者也可以通过调用unbindService()方法来停止服务，这时候Service就会调用onUnbind()－>onDestroyed()方法。

(3) 既通过startService()方式启动了Service，又通过bindService()方式启动的Service生命周期。

这种情况比较复杂。首先，Service遵循如下两个原则：

其一，"Service创建遵循唯一性原则"。即：Service的onCreate()的方法只会被调用一次，就是无论多少次调用startService()或bindService()，Service只被创建一次。

其二，"Service终止遵循二元性原则"。即：要终止该Service的终止，必须既调用了unbindService()，又调用了stopService()（二者调用先后顺序无关），才能终止Service。

然后，根据运行场景不同，运行状态说明如下：

- 如果先是bind()了，那么start()的时候就直接运行Service的onStart()方法，如果先是start()，那么bind()的时候就直接运行onBind()方法。
- 如果Service运行期间调用了bindService()，这时候再调用stopService()的话，Service是不会调用onDestroy方法的，Service就不能停止了，只能先调用UnbindService()，再调用StopService()。
- 如果一个Service通过startService()被start()之后，多次调用startService()的话，Service会多次调用onStart()方法。若多次调用stopService()，Service只会调用一

次 onDestroyed()方法。

- 如果一个 Service 通过 bindService()被 start()之后，多次调用 bindService()，Service 只会调用一次 onBind()方法。若多次调用 unbindService()，会出异常。

在什么情况下，使用 startService 或 bindService 或同时使用 startService 和 bindService 呢？

(1) 如果只是想要启动一个后台服务长期进行某项任务，那么使用 startService 便可以了。

(2) 如果你想要与正在运行的 Service 取得联系，那么有两种方法，

- 一种是使用 broadcast，该方法的缺点是如果交流较为频繁，容易造成性能上的问题，并且 BroadcastReceiver 本身执行代码的时间是很短的(最长执行时间只有 10 秒，也许执行到一半，后面的代码便不会执行)；
- 一种是使用 bindService，没有前者那些问题，因此我们肯定选择使用 bindService(这个时候你便同时在使用 startService 和 bindService 了，这在 Activity 中更新 Service 的某些运行状态是相当有用的)。

(3) 如果你的服务是通过 AIDI 定义创建，供连接上的客服端远程调用执行服务端方法。在这种情况下，建议而只用 bindService 绑定服务，这样在第一次 bindService 的时候才会创建服务的实例运行它，这会节约很多系统资源，特别是如果服务是 Remote Service，那么该效果会越明显。

Service 在 AndroidManifest. xml 的元素属性见表 8-1。

表 8-1 Service 在 AndroidManifest. xml 里的属性

属　性	含　义
android:name	服务类名
android:label	服务的名字，如果此项不设置，那么默认显示的服务名则为类名
android:icon	服务的图标
android:permission	服务的权限，这意味着只有提供了该权限的应用才能控制或连接此服务
android:process	服务所在进程，如果设置了此项，那么将服务会在包名后面加上这段字符串表示另一进程的名字
android:enabled	如果此项设置为 true，那么 Service 将会默认被系统启动，不设置默认此项为 false
android:exported	服务是否能够被其他应用程序所控制或连接，不设置默认此项为 false

8.5 Service 的类别

按照 Service 运行所在的进程来区别，Service 可以分为 Local Service 和 Remote Service 两大类。

- Local Service。该服务运行在主进程(main)线程上的。如 onCreate()和 onStart()。这些函数在被系统调用的时候都是在当前进程的主线程(main)上。如果有大量耗时操作，建议另外创建线程来执行这些代码，从而避免应用无响应(Application No Reponse，ANR)错误。

• Remote Service。该服务是运行在独立进程的主线程(main)上。

两者区别见表 8-2。

表 8-2　Local Service 与 Remote Service 的区别

Service 类别	运行载体	优　　点	缺　　点	典型应用
Local Service	当前进程的主线程	占用资源少,创建使用简单	服务生命周期与当前进程相关,当前进程结束,服务也被终止	音乐播放服务
Remote Service	独立进程的主线程	占用资源多,创建使用稍复杂	服务生命周期与当前进程无关,服务可独立运行	系统服务,如 GPS 服务

8.6　Local Service 的创建与启动

Local Service 是指所创建运行的 Service 进程 PID,与启动 Service 的组件所运行的进程 PID 相同。

【例 8-1】　教学案例设计:一个 Activity,通过一个 TextView 显示该 Activity 的 PID,通过点按该按钮启动或停止一个 Service。观察一下比较 Activity 的 PID 与 Service 的 PID 是否相同,观察 Service 启动停止生命周期所运行的函数顺序。

MainActivity 的代码如下:

```
package com.shnu.startserviceinprocess;
import android.os.Bundle;
import android.os.Process;
import android.app.Activity;
import android.content.Intent;
import android.view.View;
import android.view.View.OnClickListener;
import android.widget.Button;
import android.widget.TextView;

public class MainActivity extends Activity {
    private Button start, stop;
    private TextView info;
     private String TAG = this.getClass().getName() + "\n(pid =" + Process.myPid() + ")";

    @Override
    protected void onCreate(Bundle savedInstanceState) {
        super.onCreate(savedInstanceState);
        setContentView(R.layout.activity_main);
        start = (Button) findViewById(R.id.start);
        stop = (Button) findViewById(R.id.stop);
        info = (TextView)this.findViewById(R.id.textView1);
        info.setText(TAG);

        final Intent intent = new Intent(MainActivity.this, MyService.class);
```

```
        start. setOnClickListener(new OnClickListener() {
              @Override
              public void onClick(View arg0) {
                  startService(intent);
              }
        });
        stop. setOnClickListener(new OnClickListener() {
              @Override
              public void onClick(View arg0) {
                  stopService(intent);
              }
        });
    }

}
```

MyService 代码如下：

```
package com.shnu.startserviceinprocess;

import android.app.Service;
import android.content.Context;
import android.content.Intent;
import android.os.IBinder;
import android.os.Process;
import android.util.Log;
import android.widget.Toast;

public class MyService extends Service
{     private String TAG = this.getClass().getName() + "(pid = " + Process.myPid() + ")";
        private int i = 1;
        private String state = TAG + " -- >";
    @Override
    public IBinder onBind(Intent arg0)
    {
        tlog(this,"onBind(Intent arg0)");
        Log.i(TAG, "onBind(Intent arg0)");
        return null;
    }
    @Override
    public void onCreate()
    {
        super.onCreate();
        tlog(this,"onCreate()");
        Log.i(TAG, " onCreate()");
    }
    @Override
    public int onStartCommand(Intent intent, int flags, int startId)
    {
        tlog(this,"onStartCommand()");
```

```
        Log.i(TAG, "onStartCommand");
        return START_STICKY;
    }

    @Override
    public void onDestroy()
    {
        super.onDestroy();
        tlog(this,"onDestroy()");
        Log.i(TAG, "onDestroy()");
    }

    public void tlog(Context context,String s){
        state = state + (i++) + "." + s + " -- >";
        Toast.makeText(this, state, Toast.LENGTH_LONG).show();
        }
}
```

Activity_main.xml 布局文件内容如下：

```
<RelativeLayout xmlns:android = "http://schemas.android.com/apk/res/android"
    xmlns:tools = "http://schemas.android.com/tools"
    android:layout_width = "match_parent"
    android:layout_height = "match_parent"
    tools:context = ".MainActivity" >
    <Button
        android:id = "@ + id/start"
        android:layout_width = "wrap_content"
        android:layout_height = "wrap_content"
        android:layout_alignParentLeft = "true"
        android:layout_alignParentTop = "true"
        android:layout_marginLeft = "56dp"
        android:layout_marginTop = "126dp"
        android:text = "start" />

    <Button
        android:id = "@ + id/stop"
        android:layout_width = "wrap_content"
        android:layout_height = "wrap_content"
        android:layout_alignBaseline = "@ + id/start"
        android:layout_alignBottom = "@ + id/start"
        android:layout_marginLeft = "37dp"
        android:layout_toRightOf = "@ + id/start"
        android:text = "stop" />

    <TextView
        android:id = "@ + id/textView1"
        android:layout_width = "wrap_content"
        android:layout_height = "wrap_content"
        android:layout_alignParentTop = "true"
```

```
        android:layout_marginTop = "51dp"
        android:text = "Activity PID:" />

</RelativeLayout>
```

AndroidManifest.xml 内容如下：

```
<?xml version = "1.0" encoding = "utf - 8"?>
<manifest xmlns:android = "http://schemas.android.com/apk/res/android"
    package = "com.shnu.startserviceinprocess"
    android:versionCode = "1"
    android:versionName = "1.0" >

    <uses - sdk
        android:minSdkVersion = "8"
        android:targetSdkVersion = "17" />

    <application
        android:allowBackup = "true"
        android:icon = "@drawable/ic_launcher"
        android:label = "@string/app_name"
        android:theme = "@style/AppTheme" >
        <activity
            android:name = "com.shnu.startserviceinprocess.MainActivity"
            android:label = "@string/app_name" >
            <intent - filter >
                <action android:name = "android.intent.action.MAIN" />

                <category android:name = "android.intent.category.LAUNCHER" />
            </intent - filter >
        </activity >
        <! -- 配置一个 Service 组件 -- >
        <service android:name = ".MyService"
                android:exported = "false"
                >
            <intent - filter >
                  <! -- 为该 Service 组件的 intent - filter 配置 action -- >
                  <action android:name = "com.shnu.service.MY_SERVICE" />
            </intent - filter >
        </service >
    </application >
</manifest >
```

系统运行结果如图 8-1 和图 8-2 所示。可以看到 Activity 与 Service 同处于一个进程 PID 中，启动 Service，系统执行顺序为先调用 OnCreat()，然后调用 OnStartCommand()；停止 Service 时执行 OnDestory()方法。

Android Binder 是基于 Service 与 Client 的，有一个 ServiceManager 的守护进程管理着系统的各个服务，它负责监听是否有其他程序向其发送请求，如果有请求就响应。每个服务都要在 ServiceManager 中注册，而请求服务的客户端去 ServiceManager 请求服务。

图 8-1　Service 启动结果后回调函数执行顺序

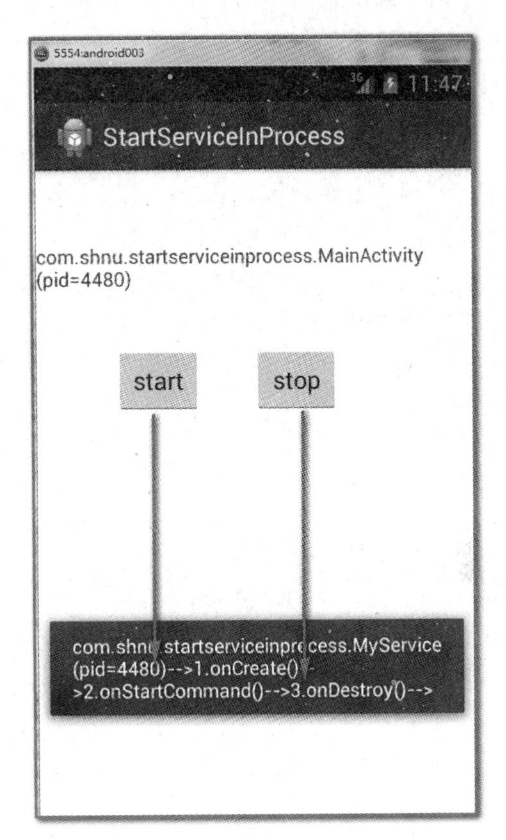

图 8-2　Service 结果回调函数执行顺序

【例 8-2】　教学案例设计：一个 Activity，通过一个 TextView 显示该 Activity 的 PID，通过单击该按钮"bind"()或"UnbindService"()解除绑定一个 Service。观察并比较 Activity 的 PID 与 Service 的 PID 是否相同，观察 Service 绑定、解除绑定过程中生命周期所运行的函数顺序。

MainActivity 的代码如下：

```
package com.shnu.bindservice;
import android.os.Bundle;
import android.os.IBinder;
import android.os.Process;
import android.view.View;
import android.view.View.OnClickListener;
import android.widget.Button;
import android.widget.TextView;
import android.app.Activity;
import android.app.Service;
import android.content.ComponentName;
import android.content.Intent;
import android.content.ServiceConnection;
```

```
public class MainActivity extends Activity {
    private Button start, stop;
    private TextView info;
     private String TAG = this.getClass().getName() + "\n(pid = " + Process.myPid() + ")";

    // 保持所启动的 Service 的 IBinder 对象
    MyService.MyBinder binder;
    // 定义一个 ServiceConnection 对象
    private ServiceConnection conn = new ServiceConnection() {
        // 当该 Activity 与 Service 连接成功时回调该方法
        @Override
        public void onServiceConnected(ComponentName name, IBinder service) {
            // 获取 Service 的 onBind 方法所返回的 MyBinder 对象
            binder = (MyService.MyBinder) service;
        }

        // 当该 Activity 与 Service 断开连接时回调该方法
        @Override
        public void onServiceDisconnected(ComponentName name) {

        }
    };
    @Override
    protected void onCreate(Bundle savedInstanceState) {
        super.onCreate(savedInstanceState);
        setContentView(R.layout.activity_main);
        start = (Button) findViewById(R.id.start);
        stop = (Button) findViewById(R.id.stop);
        info = (TextView)this.findViewById(R.id.textView1);
        info.setText(TAG);

        // 创建启动 Service 的 Intent
        final Intent intent = new Intent();
        // 为 Intent 设置 Action 属性
        intent.setAction("com.shnu.bindservice.MY_SERVICE");
        start.setOnClickListener(new OnClickListener()
        {
            @Override
            public void onClick(View arg0)
            {
                bindService(intent, conn, Service.BIND_AUTO_CREATE);
            }
        });
        stop.setOnClickListener(new OnClickListener()
        {
            @Override
            public void onClick(View arg0)
            {
                // 解除绑定 Serivce
```

```
                unbindService(conn);
            }
        });
    }
```

MyService 的代码如下：

```
package com.shnu.bindservice;

import android.app.Service;
import android.content.Context;
import android.content.Intent;
import android.os.Binder;
import android.os.IBinder;
import android.os.Process;
import android.util.Log;
import android.widget.Toast;

public class MyService extends Service
{    private String TAG = this.getClass().getName() + "(pid = " + Process.myPid() + ")";
        private int i = 1;
        private String state = TAG + " -- >";
        // 定义 onBinder 方法所返回的对象
        private MyBinder binder = new MyBinder();
        // 通过继承 Binder 来实现 IBinder 类
        public class MyBinder extends Binder
        {
        }
    @Override
    public IBinder onBind(Intent arg0)
    {
        tlog(this,"onBind(Intent arg0)");
        Log.i(TAG, "onBind(Intent arg0)");
        return binder;
    }
    @Override
    public void onCreate()
    {
        super.onCreate();
        tlog(this,"onCreate()");
        Log.i(TAG, " onCreate()");
    }
    @Override
    public int onStartCommand(Intent intent, int flags, int startId)
    {
        tlog(this,"onStartCommand()");
        Log.i(TAG, "onStartCommand");
        return START_STICKY;
    }
    // Service 被断开连接时回调该方法
```

```
        @Override
        public boolean onUnbind(Intent intent)
        {
            tlog(this,"onUnbind()");
            Log.i(TAG, "onUnbind()");
            return true;
        }
        @Override
        public void onDestroy()
        {
            super.onDestroy();
            tlog(this,"onDestroy()");
            Log.i(TAG, "onDestroy()");
        }

        public void tlog(Context context,String s){
            state = state + (i++) + "." + s + "-->";
            Toast.makeText(this, state, Toast.LENGTH_LONG).show();

        }
}
```

Activity_main.xml 布局文件内容如下：

```
<RelativeLayout xmlns:android = "http://schemas.android.com/apk/res/android"
    xmlns:tools = "http://schemas.android.com/tools"
    android:layout_width = "match_parent"
    android:layout_height = "match_parent"
    tools:context = ".MainActivity" >

    <Button
        android:id = "@+id/start"
        android:layout_width = "wrap_content"
        android:layout_height = "wrap_content"
        android:layout_alignParentTop = "true"
        android:layout_centerHorizontal = "true"
        android:layout_marginTop = "191dp"
        android:text = "Bind Service" />

    <Button
        android:id = "@+id/stop"
        android:layout_width = "wrap_content"
        android:layout_height = "wrap_content"
        android:layout_alignLeft = "@+id/start"
        android:layout_below = "@+id/start"
        android:layout_marginTop = "39dp"
        android:text = "UnBind Service" />

    <TextView
        android:id = "@+id/textView1"
```

```
        android:layout_width = "wrap_content"
        android:layout_height = "wrap_content"
        android:layout_alignParentLeft = "true"
        android:layout_alignParentTop = "true"
        android:layout_marginLeft = "16dp"
        android:layout_marginTop = "36dp"
        android:text = "TextView" />

</RelativeLayout>
```

AndroidManifest.xml 内容如下：

```
<?xml version = "1.0" encoding = "utf-8"?>
<manifest xmlns:android = "http://schemas.android.com/apk/res/android"
    package = "com.shnu.bindservice"
    android:versionCode = "1"
    android:versionName = "1.0" >

    <uses-sdk
        android:minSdkVersion = "8"
        android:targetSdkVersion = "17" />

    <application
        android:allowBackup = "true"
        android:icon = "@drawable/ic_launcher"
        android:label = "@string/app_name"
        android:theme = "@style/AppTheme" >
        <activity
            android:name = "com.shnu.bindservice.MainActivity"
            android:label = "@string/app_name" >
            <intent-filter>
                <action android:name = "android.intent.action.MAIN" />

                <category android:name = "android.intent.category.LAUNCHER" />
            </intent-filter>
        </activity>
          <!-- 配置一个 Service 组件 -->
          <service android:name = ".MyService"
                   android:exported = "false"
                   >
               <intent-filter>
                   <!-- 为该 Service 组件的 intent-filter 配置 action -->
                   <action android:name = "com.shnu.bindservice.MY_SERVICE" />
            </intent-filter>
        </service>
    </application>

</manifest>
```

运行结果如图 8-3 和图 8-4 所示。

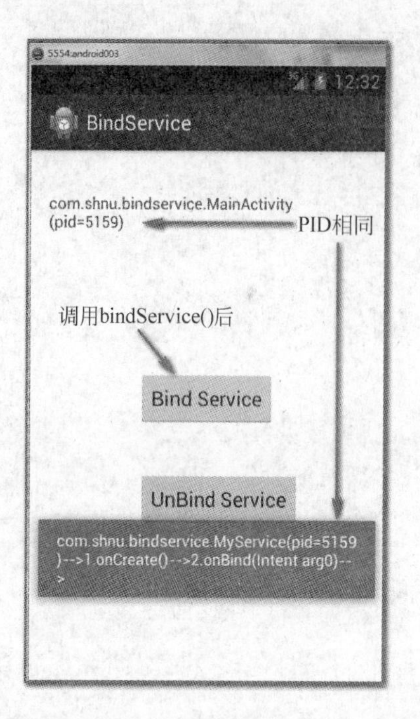

图 8-3 同一进程中绑定 Service 后回调函数顺序

图 8-4 同一进程中解除绑定 Service 后回调函数顺序

8.7 Remote Service 的创建与启动

Local Service 是指所创建运行的 Service 进程 PID，与启动 Service 的组件所运行的进程 PID 不同。

【例 8-3】 教学案例设计：在一个 Application 中，建立一个 Activity，该 Activity 上一个 TextView 显示该 Activity 的 PID，通过单击该按钮启动或停止另一个进程【例 8-1】中的 Service。观察并比较 Activity 的 PID 与 Service 的 PID 是否相同，观察 Service 启动停止生命周期所运行的函数顺序。

MainActivity 的代码如下：

```
package com.shnu.startserviceindifferentprocess;
import android.os.Bundle;
import android.os.Process;
import android.app.Activity;
import android.content.Intent;
import android.view.View;
import android.view.View.OnClickListener;
import android.widget.Button;
import android.widget.TextView;

public class MainActivity extends Activity {
    private Button start, stop;
    private TextView info;
```

```
    private String TAG = this.getClass().getName() + "\n(pid = "
                + Process.myPid() + ")";

    @Override
    protected void onCreate(Bundle savedInstanceState) {
        super.onCreate(savedInstanceState);
        setContentView(R.layout.activity_main);
        start = (Button) findViewById(R.id.start);
        stop = (Button) findViewById(R.id.stop);
        info = (TextView) this.findViewById(R.id.textView1);
        info.setText(TAG);

        final Intent intent = new Intent();
        intent.setAction("com.shnu.service.MY_SERVICE");
        start.setOnClickListener(new OnClickListener() {

            @Override
            public void onClick(View arg0) {
                startService(intent);
            }
        });
        stop.setOnClickListener(new OnClickListener() {

            @Override
            public void onClick(View arg0) {
                stopService(intent);
            }
        });

    }

}
```

AndroidManifest.xml 内容如下：

```
<?xml version = "1.0" encoding = "utf - 8"?>
<manifest xmlns:android = "http://schemas.android.com/apk/res/android"
    package = "com.shnu.startserviceindifferentprocess"
    android:versionCode = "1"
    android:versionName = "1.0" >

    <uses - sdk
        android:minSdkVersion = "8"
        android:targetSdkVersion = "17" />

    <application
        android:allowBackup = "true"
        android:icon = "@drawable/ic_launcher"
        android:label = "@string/app_name"
        android:theme = "@style/AppTheme" >
```

```
            <activity
                android:name = "com.shnu.startserviceindifferentprocess.MainActivity"
                android:label = "@string/app_name" >
                <intent - filter>
                    <action android:name = "android.intent.action.MAIN" />

                    <category android:name = "android.intent.category.LAUNCHER" />
                </intent - filter>
            </activity>
        </application>

    </manifest>
```

另外,需要把【例 8-1】中的 AndrManifest.xml,有关 Service 的属性设置更改为:

```
<service android:name = ".MyService" android:exported = "true" >
```

即允许该 Service 暴露给其他进程使用。系统运行结果见图 8-5 和图 8-6。

图 8-5 不同进程中启动 Service 回调函数执行

图 8-6 不同进程中停止 Service 回调函数执行顺序

【例 8-4】 教学案例设计:在一个 Application 中,建立一个 Activity,该 Activity 上一个 TextView 显示该 Activity 的 PID,通过单击该按钮 bind()或 UnbindService 解除绑定另一个进程【例 8-2】中的 Service。观察并比较 Activity 的 PID 与 Service 的 PID 是否相同,观察 Service 启动停止生命周期所运行的函数顺序。

MainActivity 的代码如下:

```
package com.shnu.bindmyserviceindifferentprocess;
import android.os.Bundle;
```

```
import android.os.IBinder;
import android.os.Process;
import android.view.View;
import android.view.View.OnClickListener;
import android.widget.Button;
import android.widget.TextView;
import android.app.Activity;
import android.app.Service;
import android.content.ComponentName;
import android.content.Intent;
import android.content.ServiceConnection;
    /*
     * 不同进程的绑定请参照上例的 AIDL
     */
public class MainActivity extends Activity {
    private Button start, stop;
    private TextView info;
    private String TAG = this.getClass().getName() + "\n(pid = " + Process.myPid() + ")";

    // 保持所启动的 Service 的 IBinder 对象
    //Service binder;
    // 定义一个 ServiceConnection 对象
    private ServiceConnection conn = new ServiceConnection() {
        // 当该 Activity 与 Service 连接成功时回调该方法
        @Override
        public void onServiceConnected(ComponentName name, IBinder service) {
            // 获取 Service 的 onBind 方法所返回的 MyBinder 对象
            //binder = (Service) service;
        }

        // 当该 Activity 与 Service 断开连接时回调该方法
        @Override
        public void onServiceDisconnected(ComponentName name) {

        }
    };
    @Override
    protected void onCreate(Bundle savedInstanceState) {
        super.onCreate(savedInstanceState);
        setContentView(R.layout.activity_main);
        start = (Button) findViewById(R.id.start);
        stop = (Button) findViewById(R.id.stop);
        info = (TextView)this.findViewById(R.id.textView1);
        info.setText(TAG);

        // 创建启动 Service 的 Intent
        final Intent intent = new Intent();
        // 为 Intent 设置 Action 属性
        intent.setAction("com.shnu.bindservice.MY_SERVICE");
        start.setOnClickListener(new OnClickListener()
```

```
        {
            @Override
            public void onClick(View arg0)
            {
                bindService(intent, conn, Service.BIND_AUTO_CREATE);
            }
        });
        stop.setOnClickListener(new OnClickListener()
        {
            @Override
            public void onClick(View arg0)
            {
                // 解除绑定 Serivce
                unbindService(conn);
            }
        });
    }

}
```

AndroidManifest.xml 内容如下：

```
<?xml version = "1.0" encoding = "utf - 8"?>
<manifest xmlns:android = "http://schemas.android.com/apk/res/android"
    package = "com.shnu.bindmyserviceindifferentprocess"
    android:versionCode = "1"
    android:versionName = "1.0" >

    <uses - sdk
        android:minSdkVersion = "8"
        android:targetSdkVersion = "17" />

    <application
        android:allowBackup = "true"
        android:icon = "@drawable/ic_launcher"
        android:label = "@string/app_name"
        android:theme = "@style/AppTheme" >
        <activity
            android:name = "com.shnu.bindmyserviceindifferentprocess.MainActivity"
            android:label = "@string/app_name" >
            <intent - filter >
                <action android:name = "android.intent.action.MAIN" />

                <category android:name = "android.intent.category.LAUNCHER" />
            </intent - filter >
        </activity >
    </application >

</manifest >
```

另外，需要把【例 8-1】中的 AndroidManifest. xml，有关 Service 的属性设置更改为：

```
<service android:name = ".MyService" android:exported = "true" >
```

即允许该 Service 暴露给其他进程使用。

运行结果如图 8-7 和图 8-8 所示。

图 8-7　跨进程绑定 Service 后回调函数执行顺序

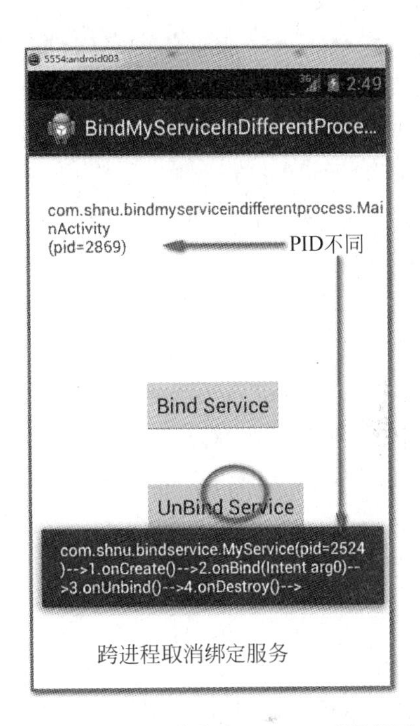

图 8-8　跨进程解除绑定 Service 后回调函数执行顺序

8.8　AIDL 与跨进程服务调用

由于 Android 每一个应用程序都运行在自己的进程(PID)空间内，当然一个 Android 应用所包含的不同组件可能运行在不同的进程(PID)中。比如，一个 Android 应用中的 UI 进程和 PID Service 进程 PID 可以不相同。Android 系统的安全性规定，一个进程通常不能访问另一个进程的内存空间，不同进程之间的对象传递，首先需要将要传递对象分解成操作系统可以理解的原始基本数据类型，然后引领(marshall)这些对象实现跨进程访问。Android 提供了 AIDL(Android Interface Definition Language，Android 接口描述语言)工具来处理这项工作。

AIDL 是一种语言，主要用来定义进程之间的接口，两个进程通过实现该接口，就可实现进程之间的交互对象传递。如，一个进程中访问另一个进程中某个对象的方法。使用 AIDL 实现 IPC(Implementing IPC Using AIDL)的机制是基于接口的，在某些程度上与 COM 或 Corba 类似，但它是轻量级的。它使用代理类在客户端和服务端之间传递值。

服务端：使用 AIDL 实现 IPC，提供业务逻辑。

客户端：调用一个 AIDL(IPC)类本地，获取服务端业务逻辑服务。

AIDL 实现 IPC 服务与绑定的步骤如下：

(1) 创建服务端和客户端进程通信可用 AIDI 接口文件。该文件(MyInterface. aidl)定义了客户端可用的方法和数据的接口。类似于 Java 的 Interface 文件。

(2) 在 Eclipse+ADT 环境中，AIDI 文件(MyInterface. aidl)会自动在 gen 目录下生成一个与 AIDI 文件名对应的 Java Interface 文件(MyInterface. java)。该接口有一个继承的命名为 MyInterface. Stub 的静态内部抽象类(并且实现了一些 IPC 调用的附加方法)。

(3) 服务端要做的工作是：继承 Service 并且重载 Service. onBind(Intent)以返回一个继承于 MyInterface. Stub 的类并且实现在. aidl 文件中声明的方法。

(4) 客户端要做的工作是：实现 ServiceConnection，调用 Context. bindService()，传递 ServiceConnection 的实现，在 ServiceConnection. onServiceConnected()方法中会接收到 IBinder 对象，调用 MyInterfaceName. Stub. asInterface(service)将返回值转换为 MyInterfaced 对象类型。即可访问服务器端方法或数据。

【例 8-5】 教学案例设计：在一个 Application 中，建立一个 Activity，该 Activity 上一个 TextView 显示该 Activity 的 PID，通过单击该按钮 bind 绑定或 UnbindService 解除绑定另一个进程中的 Service。观察并比较 Activity 的 PID 与 Service 的 PID 是否相同，观察 Service 启动停止生命周期所运行的函数顺序。

Aidl 接口定义——IstringData. aidl 的代码如下：

```
package com.shnu.startaidlservice;
interface IStringData
{
    String getString();
}
```

服务端 AidlService .java 的代码如下：

```
package com.shnu.startaidlservice;
import com.shnu.startaidlservice.IStringData.Stub;
import android.app.Service;
import android.content.Intent;
import android.os.IBinder;
import android.os.RemoteException;

public class AidlService extends Service
{   //上一案例已经演示生命周期,故省略
    private StringBinder stringBinder;
    public class StringBinder extends Stub
    {
        @Override
        public String getString() throws RemoteException
        {
            return "这是 AidlService 返回的字符串";
        }
    }
```

```
    @Override
    public void onCreate()
    {
        super.onCreate();
        stringBinder = new StringBinder();
    }
    @Override
    public IBinder onBind(Intent arg0)
    {
        return stringBinder;
    }
    @Override
    public void onDestroy()
    {
    }
}
```

在服务器端,AndroidManifest.xml 文件内容如下：

```
<?xml version = "1.0" encoding = "utf - 8"?>
<manifest xmlns:android = "http://schemas.android.com/apk/res/android"
    package = "com.shnu.startaidlservice"
    android:versionCode = "1"
    android:versionName = "1.0" >

    <uses - sdk
        android:minSdkVersion = "8"
        android:targetSdkVersion = "17" />

    <application
        android:icon = "@drawable/ic_launcher"
        android:label = "@string/app_name"
         >
        <!-- 定义一个 Service 组件 -->
        <service android:name = ".AidlService"
          >
            <intent - filter>
                    <action android:name = "com.shnu.startaidlservice.AIDL_SERVICE" />
            </intent - filter>
        </service>
    </application>

</manifest>
```

客户端 MainActivity.java 的代码如下：

```
package com.shnu.startaidlclient;

import com.shnu.startaidlservice.IStringData;
```

```
import android.os.Bundle;
import android.os.IBinder;
import android.os.RemoteException;
import android.app.Activity;
import android.app.Service;
import android.content.ComponentName;
import android.content.Intent;
import android.content.ServiceConnection;
import android.view.View;
import android.view.View.OnClickListener;
import android.widget.Button;
import android.widget.TextView;

public class MainActivity extends Activity {
    private Button get;
    private TextView result;
    private IStringData iStringData;
    private ServiceConnection conn = new ServiceConnection()
    {
        @Override
        public void onServiceConnected(ComponentName name, IBinder service)
        {
            // 获取远程 Service 的 onBind 方法返回的对象的代理
            iStringData = IStringData.Stub.asInterface(service);
        }

        @Override
        public void onServiceDisconnected(ComponentName name)
        {
            iStringData = null;
        }
    };
    @Override
    protected void onCreate(Bundle savedInstanceState) {
        super.onCreate(savedInstanceState);
        setContentView(R.layout.activity_main);
        get = (Button) findViewById(R.id.get);
        result = (TextView) findViewById(R.id.result);
        // 创建所需绑定服务的 Intent
        final Intent intent = new Intent();
        intent.setAction("com.shnu.startaidlservice.AIDL_SERVICE");
        bindService(intent, conn, Service.BIND_AUTO_CREATE);
        get.setOnClickListener(new OnClickListener()
        {
            @Override
            public void onClick(View arg0)
            {
                try {
```

```
                    result.setText(iStringData.getString());
                } catch (RemoteException e) {
                    // TODO Auto - generated catch block
                    e.printStackTrace();
                }
            }
        });
    }
    @Override
    public void onDestroy()
    {
        super.onDestroy();
        // 解除绑定
        this.unbindService(conn);
    }

}
```

系统运行结果如图 8-9 所示。

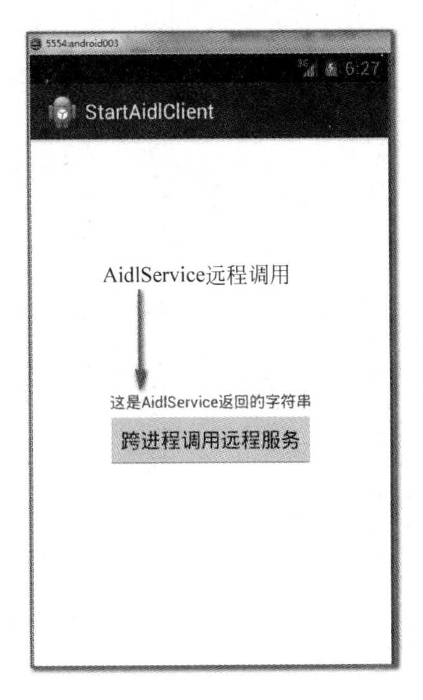

图 8-9　跨进程调用远程服务

8.9　本章小结

通过本章学习,我们已经掌握了 Android 系统的 Service 基础知识,掌握了启动、绑定、同时启动与绑定 Service 方法,以及对应的 Service 生命周期。掌握了进程内与进程间启动与绑定 Service 的方法,掌握了基本的 AIDL 服务端与客户端程序编写方法。

8.10 习题与课外阅读

8.10.1 习题

(1) 分析 startService()方法启动的一个 Service 的回调函数执行顺序。

(2) 分析 bindService()方法绑定的 Service 的回调函数执行顺序。

(3) 比较一下 AIDL 与 Java Interface 接口文件异同。

(4) Service 与 Thread 有哪些不同?

8.10.2 课外阅读

访问下列技术网站,了解一下 AIDL 传送对象,Service 与客户端程序编写:

- http://www.android.com;
- http://source.android.com;
- http://www.openhandsetalliance.com/。

第9章 Android文件及数据库

几乎所有的Android应用程序,都涉及对应用程序的初始化信息、运行状态、数据等数据的保存,这些操作会涉及Android的文件或数据库,本章主要介绍Android系统文件与数据库的读写基本知识。

本章学习目标:

- 掌握Android系统文件安全模型概念;
- 掌握Android应用程序中对文件的读写方法;
- 掌握SQLite数据库创建、数据读、写、改、查等方法。

9.1 Android系统文件安全模型

Android系统有一套独特的系统安全模型,当一个应用程序(.apk)在安装时,系统会自动为该应用和应用本身的资源文件分配一个用户userid,通常一个Android应用对应一个用户userid。

在默认情况下,一个Android应用只能在程序所在的userid空间内读写文件,或创建数据库、对数据库进行增删查改操作,而userid权限之外的其他程序无此权限。

这种机制保证了Android读写文件的安全,一个进程打开文件时,Android系统会查验进程的userid。所以通常不能直接用Java的API的文件来读写文件,因为Java的IO函数没有提供userid安全机制。

9.2 资源文件的访问

一个Android应用程序的运行需要许多资源文件的支持,比如,图标文件、数据文件、配置文件等,这些文件随Android应用程序编译后代码一起打包成一个APK文件发布。这类文件是APK包内部私有文件,属于静态文件,这些资源文件不允许其他应用修改(如软件用户版权信息),只允许应用程序本身来访问。这类文件按照文件存放位置不同,可以分为两类资源文件:

(1) 保存resource中的/res/raw目录下的原始数据文件;

(2) 保存在assets目录下的原始数据文件。

如果资源文件是文本文件,文件的读取与显示则需要考虑文件的编码和换行符。如果在中文环境下使用Eclipse,编辑好的文本文件默认使用GBK编码格式。建议用户把编辑器默认编码改成UTF-8格式(见图9-1)。

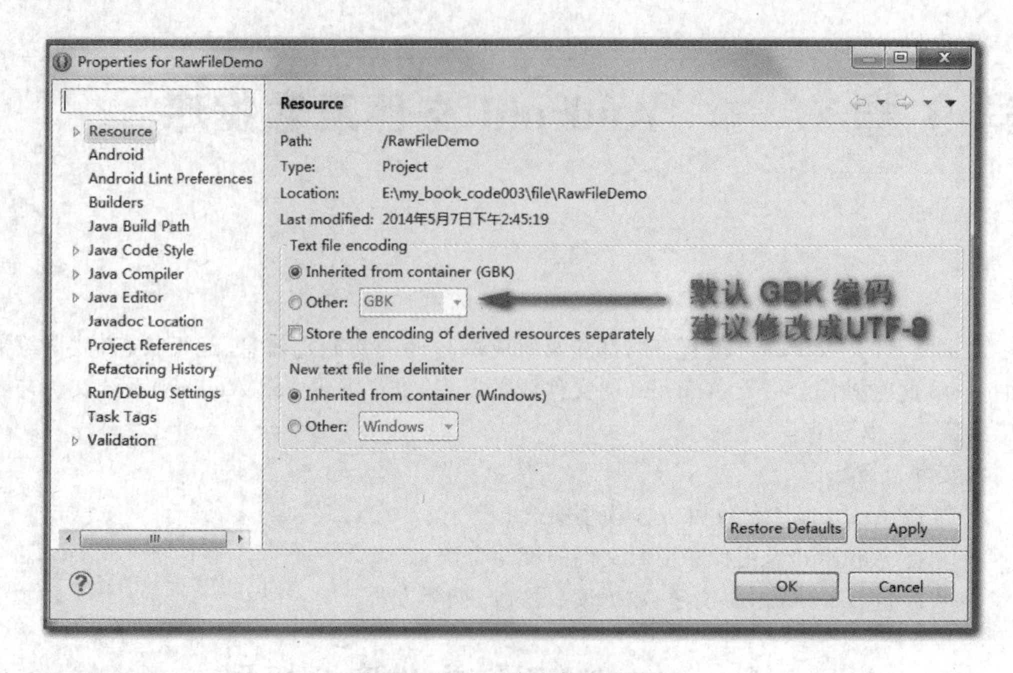

图 9-1　Eclipse 中文件编码设定

9.2.1　/res/raw 目录下的原始数据文件的访问

Android 系统中每个应用程序的资源目录/res/raw 下的文件，称为原始文件，Android 为该目录下的文件生成唯一资源 ID。

通过 this. getResources(). openRawResource(R. raw. rawfile)函数调用(其中 rawfile 是/res/raw 下的文件 ID)；返回文件的输入流为 InputStream，有了输入流就可以实现对文件内容的访问。由于只能获取该文件的输入流，因此不能对该文件进行写操作。

【例 9-1】 教学案例设计：设计一个程序，读取/res/raw 目录下文件的内容。

首先设计一个工具函数 inputStream2String(InputStream in)，将一个输入流转换成字符串，主要代码如下：

```
public String inputStream2String(InputStream in) {
        String content = "";
        try {
            int len = in.available();
            byte[] buffer = new byte[len];
            in.read(buffer);
            content = new String(buffer, "gbk");

        } catch (Exception e) {

            e.printStackTrace();
        }
        return content;

    }
```

用于显示/res/raw 目录下的原始数据文件内容的主要实现代码如下：

```
public void showRawFile(View view) {
        InputStream fins = getResources().openRawResource(R.raw.rawfile);
        String content = this.inputStream2String(fins);
        tv.setText(this.getResources().getResourceName(R.raw.rawfile));
        et.setText(content);

    }
```

9.2.2 /assets 目录下的原始数据文件的访问

Android 系统为每个应用程序提供了/assets 目录，/assets 目录下的文件称为原生文件，这类文件在被打包成 apk 文件时是不会进行压缩的。保存在/res 和/assets 目录下文件的不同点是，Android 系统不为/assets 下的文件生成资源 ID。如果使用/assets 下的文件，需要指定文件的路径和文件名。

Android 系统使用 AssetManager 对/assets 目录下文件进行访问，通过 getResources().getAssets()获得 AssetManager，然后使用 open（"文件名"）方法获取该文件的输入流 InputStream；使用 Context. getAssets(). open("sample. txt")可以获取文件输入流，然后用户可以按照流的方式读取文件内容。

【例 9-2】 教学案例设计：设计一个程序，读取/assets 目录下的原始数据文件的内容。

主要代码如下：

```
public void showAssetsFile(View view) {

        try {
            String fileName = "as.txt";
            tv.setText(fileName);
            InputStream fins = getAssets().open(fileName);
            String content = this.inputStream2String(fins);
            et.setText(content);

        } catch (Exception e) {

            e.printStackTrace();
        }
    }
```

9.3 Android 设备内部存储文件的读写

Android 系统提供了设备内部存储文件的读与写操作 API，可以非常方便地完成文件的创建、追加文件内容、读取文件内容等操作。文件默认存放在"./data/data/包名/files/"目录下。

对于 API 17 以上版本，主要有两种文件操作模式：

- MODE_PRIVATE——默认，创建或替换该文件名的文件作为应用程序私有文件。

- MODE_APPEND——在已有的文件后面追加内容,如果文件不存在,则先创建文件再追加内容。
- 文件输入流获取:context.openFileInput()返回 FileInputStream 读文件。
- 文件输出流获取:context.openFileOutput()返回 FileOutputStream 写文件。

【例 9-3】 教学案例设计:设计一个程序,演示读写 Android 设备内部存储设备中的文件。

程序示例代码如下:

```
String FILENAME = "hello_file";
String string = "hello world!";
FileOutputStream fos = openFileOutput(FILENAME, Context.MODE_PRIVATE);
fos.write(string.getBytes());
fos.close();
```

9.4 Android 外部存储设备文件的读写

9.4.1 外部存储设备检测

相对于 Android 内部设备中的文件的读写,Android 外部存储设备文件的读写运行环境稍微复杂一点,因为外部设备(通常是 SD 卡)有可能在使用过程中卸载。

通常在对 Android 外部存储设备文件读写之前,要使用函数 Environment.getExternalStorageState()检查外部存储设备的状态。

```
boolean mExternalStorageAvailable = false;
boolean mExternalStorageWriteable = false;
String state = Environment.getExternalStorageState();
if (Environment.MEDIA_MOUNTED.equals(state)) {
    // 设备可读写
    mExternalStorageAvailable = mExternalStorageWriteable = true;
} else if (Environment.MEDIA_MOUNTED_READ_ONLY.equals(state)) {
    // 设备只读
    mExternalStorageAvailable = true;
    mExternalStorageWriteable = false;
} else {
    // 设备不可读写
    mExternalStorageAvailable = mExternalStorageWriteable = false;
}
```

9.4.2 外部存储设备上私有文件读写

在 API 8 以上版本,使用 getExternalFilesDir()打开一个外部存储文件目录,这个方法需传递一个指定类型目录的参数,如 DIRECTORY_MUSIC 和 DIRECTORY_RINGTONES,若为空,则返回应用程序文件根目录;此方法可以根据指定的目录类型创建目录,该文件为应用程序私有,如果卸载应用程序,目录将会一起删除。

在 API 7 以下版本，使用 getExternalStorageDirectory()打开外部存储文件根目录，文件写入到如下目录中：

```
/Android/data/<package_name>/files/
```

【例 9-4】 教学案例设计：设计一个程序，演示外部存储设备上私有文件读写。

```
void createExternalStoragePrivateFile() {
    // 在外部存储设备上创建私有文件
    File file = new File(getExternalFilesDir(null), "DemoFile.jpg");

    try {

        InputStream is = getResources().openRawResource(R.drawable.balloons);
        OutputStream os = new FileOutputStream(file);
        byte[] data = new byte[is.available()];
        is.read(data);
        os.write(data);
        is.close();
        os.close();
    } catch (IOException e) {
        // 无法创建文件,因外部存储设备未装载
        Log.w("ExternalStorage", "Error writing " + file, e);
    }
}

void deleteExternalStoragePrivateFile() {
    // 取得外部存储设备路径
    File file = new File(getExternalFilesDir(null), "DemoFile.jpg");
    if (file != null) {
        file.delete();
    }
}

boolean hasExternalStoragePrivateFile() {
    // 取得外部存储设备路径
    File file = new File(getExternalFilesDir(null), "DemoFile.jpg");
    if (file != null) {
        return file.exists();
    }
    return false;
}
```

9.4.3 外部存储设备上公有文件读写

如果想将文件保存为不为应用程序私有，在应用程序卸载时不被删除，需要将文件保存到外部存储的公共目录上，这些目录在存储设备根目录下，如音乐、图片，铃声等。

在 API 8 以上版本，使用 getExternalStoragePublicDirectory()传入一个公共目录的类型参数，如 DIRECTORY_MUSIC、DIRECTORY_PICTURES、DIRECTORY_RINGTONES 等，目录

不存在时这个方法会为你创建目录。

在 API 7 以下版本中，使用 getExternalStorageDirectory()打开存储文件根目录，保存文件到下面的目录中：

/mnt/sdcard/Music

/mnt/sdcard/Podcasts

/mnt/sdcard/Ringtones

/mnt/sdcard/Alanns

/mnt/sdcard/Notification

/mnt/sdcard/Pictures

/mnt/sdcard/Movies

/mnt/sdcard/Downloads

【例 9-5】 教学案例设计：设计一个程序，演示外部存储设备上公有文件读写。

```
void createExternalStoragePublicPicture() {
    // 在用户公有图片目录下取得存放图片路径
    File path = Environment.getExternalStoragePublicDirectory(
            Environment.DIRECTORY_PICTURES);
    File file = new File(path, "DemoPicture.jpg");

    try {
        // 确认目录存在
        path.mkdirs();
        InputStream is = getResources().openRawResource(R.drawable.balloons);
        OutputStream os = new FileOutputStream(file);
        byte[] data = new byte[is.available()];
        is.read(data);
        os.write(data);
        is.close();
        os.close();
    } catch (IOException e) {
         Log.w("ExternalStorage", "Error writing " + file, e);
    }
}

void deleteExternalStoragePublicPicture() {
    // 删除外部存储介质上的图片文件
    File path = Environment.getExternalStoragePublicDirectory(
            Environment.DIRECTORY_PICTURES);
    File file = new File(path, "DemoPicture.jpg");
    file.delete();
}

boolean hasExternalStoragePublicPicture() {
    // 判断外部存储介质上是否有 DemoPictuwre.jpg 文件存在
    File path = Environment.getExternalStoragePublicDirectory(
            Environment.DIRECTORY_PICTURES);
    File file = new File(path, "DemoPicture.jpg");
    return file.exists();
}
```

9.5 Shared Preferences 文件读写

Android 平台给我们提供了一个 SharedPreferences 类，它是一个轻量级的存储类，特别适合用于保存软件配置参数。SharedPreferences 可用于存取和修改软件配置参数数据的接口存放键值对，保存用户个人首选项信息，如喜爱音乐、主题、界面风格等初始化参数。类似于 Windows 软件通常采用 ini 文件进行保存软件配置数据，或者 Java SE 应用采用 properties 属性文件进行保存软件配置数据。Android 使用 SharedPreferences 保存数据，所保存的文件是 xml 文件，文件存放在/data/data 目录中。

下面介绍使用 getSharedPrefernces()的步骤。

9.5.1 写操作

(1) 获得 SharedPreferences 的实例对象，通过 getSharedPreferences()传递文件名和模式；

(2) 获得 Editor 的实例对象，通过 SharedPreferences 的实例对象的 edit()方法；

(3) 存入数据，利用 Editor 对象的 putXXX()方法；

(4) 提交修改的数据，利用 Editor 对象的 commit()方法。

核心代码如下：

```
SharedPreferences sharedPreferences = getSharedPreferences("SharedPreferencesDemo", Context.MODE
_PRIVATE);
Editor editor = sharedPreferences.edit();          //获取编辑器
    editor.putString("name", "James");
    editor.putInt("age", 45);
    editor.commit();                               //提交修改
```

生成的 SharedPreferencesDemo.xml 文件内容如下：

```
<?xml version = '1.0' encoding = 'utf - 8' standalone = 'yes' ?>
< map >
    < string name = "name"> James </string >
    < int name = "age" value = "40" />
</map >
```

9.5.2 读操作

(1) 获得 SharedPreferences 的实例对象，通过 getSharedPreferences()传递文件名和模式；

(2) 读取数据，通过 SharedPreferences 的实例对象的 getXXX()方法。

```
SharedPreferences sharedPreferences = getSharedPreferences ( " SharedPreferencesDemo ",
Context.MODE_PRIVATE);
//getString()第二个参数为默认值，如果 preference 中不存在该 key，将返回默认值
String name = sharedPreferences.getString("name", "");
int age = sharedPreferences.getInt("age", 1);
```

9.6 SQLite 数据库

SQLite 是 D. Richard Hipp 用 C 语言编写的开源嵌入式数据库引擎，它是一个进程内库，实现了一个自包含的、无服务器、零配置、事务性的 SQL 数据库引擎。SQLite 由以下几个部分组成：SQL 编译器、内核、后端以及附件。所有 SQL 语句都被编译成可以在 SQLite 虚拟机中执行的程序集。SQLite 没有一个独立的服务器进程。

SQLite 数据库直接以普通的磁盘文件读写，数据库文件格式是跨平台的。SQLite 没有数据库用户权限概念，而是根据操作系统登录用户所拥有的文件系统权限来确定所有数据库的权限。每个数据库都是以单个文件的形式存在。数据库所使用的多个表、索引、触发器和视图的完整 SQL 数据库，都包含在一个单一的磁盘文件中，这些数据都是以 B-Tree 的数据结构形式存储在磁盘上的。SQLite 可以支持高达 2TB 大小的数据库。

目前没有可用于 SQLite 的网络服务器。从应用程序运行位于其他计算机上的 SQLite 的唯一方法是以网络共享方式运行。这样会导致一些问题，像 UNIX 和 Windows 网络共享都存在文件锁定问题。还有由于与访问网络共享相关的延迟而带来的性能下降问题。SQLite 只提供数据库级的锁定。在事务处理方面，SQLite 通过数据库级上的独占性和共享锁来实现独立事务处理。这意味着多个进程可以在同一时间从同一数据库读取数据，但只有一个进程可以写入数据。某个进程或线程执行数据库写操作之前，必须获得独占锁。在获得独占锁之后，其他的读或写操作将不会再发生。

SQLite 采用动态数据类型，当某个值插入到数据库时，SQLite 将会检查它的类型，如果该类型与关联的列不匹配，SQLite 则会尝试将该值转换成该列的类型，如果不能转换，则该值将作为本身的类型存储，SQLite 称这为“弱类型”。SQLite 支持 NULL、INTEGER、REAL、TEXT 和 BLOB 数据类型，分别代表空值、整型值、浮点值、字符串文本、二进制对象。

此外，SQLite 不支持某些标准的 SQL 功能，特别是外键约束（FOREIGN KEY constrains），嵌套 transcaction 和 RIGHT OUTER JOIN 和 FULL OUTER JOIN，还有一些 ALTER TABLE 功能。

除了上述功能外，SQLite 是一个完整的 SQL 系统，拥有完整的触发器、交易等。

9.6.1 SQLiteOpenHelper 类

Android 系统集成了 SQLite 系统，并提供了非常简单的 API 来创建、使用 SQLite 数据库，利用这些 API 每个 Android 应用程序都可以使用 SQLite 数据库。

Android 提供了 SQLiteOpenHelper 类来创建一个数据库，应用程序创建的数据库存储在“data/＜项目文件夹＞/databases/”目录下。SQLiteOpenHelper，封装了创建和更新数据库使用的逻辑。这个类是 SQLite 数据库的创建和版本管理类，提供了一些函数，使得程序员可以方便地对数据库进行创建、打开和版本的管理，仅有 3 个抽象方法需要实现：

- SQLiteOpenHelper (Context context, String name, SQLiteDatabase. CursorFactory factory, int version)。

- public abstract void onCreate(SQLiteDatabase db)。
- public abstract void onUpgrade(SQLiteDatabase db, int oldVersion, int newVersion)。

1. 构造函数

构造函数的语法格式如下：

```
SQLiteOpenHelper(Context context, String name, SQLiteDatabase.CursorFactory factory, int version)
```

函数相关参数：

- Context context——调用该函数的应用上下文(如 Activity)；
- String name——数据库的名称(如果为 null,数据库为一个内存中的数据库)；
- SQLiteDatabase. CursorFactory factory——记录游标对象；
- int version——数据库的版本号,该版本号是你开发的自定义版本的编号。

这个构造函数是必须有的。

2. void onCreate (SQLiteDatabase db)回调函数

该函数是一个回调函数,具体来说,就是"当这个数据库被创建的"的时候,也就是之前没有这个数据库,那么在 new SQLiteOpenHelper()的时候,就会回调这个 onCreate()函数,但是当系统中已经存在这个数据库时 new SQLiteOpenHelper(),系统就不会调用这个函数。

3. void onUpgrade (SQLiteDatabase db, int oldVersion, int newVersion)回调函数

当版本号发生变化的时候,系统就会回调这个函数,可以在这个函数中写入你想在更新数据库版本时的操作,注意：是更新数据库版本,不是简单地更新数据库中的数据！

下面的示例代码展示了如何继承 SQLiteOpenHelper 创建数据库：

```
public class DatabaseHelper extends SQLiteOpenHelper {
  DatabaseHelper(Context context, String name, CursorFactory cursorFactory, int version)
  {
    super(context, name, cursorFactory, version);
    }

    @Override
    public void onCreate(SQLiteDatabase db) {
        // TODO 创建数据库后,对数据库的操作
    }

    @Override
public void onUpgrade(SQLiteDatabase db, int oldVersion, int newVersion) {
        // TODO 更改数据库版本的操作
    }

@Override
public void onOpen(SQLiteDatabase db) {
        super.onOpen(db);
        // TODO 每次成功打开数据库后首先被执行
    }
}
```

9.6.2 SQLDatabase 类

这个类主要是对 SQLite 数据库中的数据进行管理的类，可以完成数据的增、删、查、改等操作的类。

- openDatabase()：打开数据库。
- openOrCreateDatabase()：如果数据库已经存在就打开；若不存在，就先创建再打开。如：

```
SQLiteDatabase db = openOrCreateDatabase("test.db", Context.MODE_PRIVATE, null);
```

在执行完上面的代码后，系统就会在/data/data/[PACKAGE_NAME]/databases 目录下生成一个 test.db 的数据库文件。

- execSQL(String sql)：接收一个 SQL 语句，并执行。
- 常用的查询函数有：

```
db.rawQuery(String sql, String[] selectionArgs);
db.query(String table, String[] columns, String selection, String[] selectionArgs, String
groupBy, String having, String orderBy);
db.query(String table, String[] columns, String selection, String[] selectionArgs, String
groupBy, String having, String orderBy, String limit);
db.query(String distinct, String table, String[] columns, String selection, String[]
selectionArgs, String groupBy, String having, String orderBy, String limit);
```

参数说明：

table——数据库表的名称。

columns——数据库列名称数组写入后最后返回的 Cursor 中只能查到这里的列的内容。

selection——查询条件。

selectionArgs——查询结果。

groupBy——分组列。

having——分组条件。

orderBy——排序列。

limit——分页查询限制。

查询结果返回一个 Cursor 对象。Cursor 是一个游标接口，每次查询的结果都会保存在 Cursor 中，可以通过遍历 Cursor 的方法拿到当前查询到的所有信息。Cursor 常用的函数有：

```
close()                                          //关闭游标,释放资源
copyStringToBuffer(int columnIndex, CharArrayBuffer buffer)     //在缓冲区中检索请求的列的
                                                                //文本,将将其存储
getColumnCount()                                 //返回所有列的总数
move(int offset);                                //以当前位置为参考,移动到指定行
```

```
moveToFirst();                          //移动到第一行
moveToLast();                           //移动到最后一行
moveToPosition(int position);           //移动到指定行
moveToPrevious();                       //移动到前一行
moveToNext();                           //移动到下一行
isFirst();                              //是否指向第一条
isLast();                               //是否指向最后一条
isBeforeFirst();                        //是否指向第一条之前
isAfterLast();                          //是否指向最后一条之后
isNull(int columnIndex);                //指定列是否为空(列基数为 0)
isClosed();                             //游标是否已关闭
getCount();                             //总数据项数
getPosition();                          //返回当前游标所指向的行数
getColumnIndex(String columnName);      //返回某列名对应的列索引值
getString(int columnIndex);             //返回当前行指定列的值
```

虽然 SQLite 使用 execSQL(String sql)执行 SQL 语句可以实现数据的增、删、查、改等操作,但是为了更加方便,开发者系统也提供了一些函数进行数据库的访问,常用的函数如下:

- insert()——添加数据;
- update()——更新数据;
- delete()——删除数据;
- query()——查询数据。

9.6.3 SQLite 数据库管理工具

使用 Android 模拟器,有两种可供选择的方法来管理数据库。

其一,模拟器绑定了 sqlite3 控制台程序,可以使用 adb shell 命令来调用它。只要进入了模拟器的 shell,在数据库的路径执行 sqlite3 命令就可以了。数据库文件一般存放在:

```
/data/data/your.app.package/databases/your-db-name
```

其二,如果你喜欢使用更友好的工具,可以把数据库复制到开发机上,使用 SQLite-aware 客户端来操作它。这样就可在一个数据库的副本上操作,如果想要你的修改能反映到设备上,复制需要把数据库备份回去。

把数据库从设备上复制出来,可以使用 adb pull 命令(或者在 IDE 上做相应操作)。存储一个修改过的数据库到设备上,使用 adb push 命令。

9.6.4 数据库综合应用示例

【例 9-6】 使用 SQLite 数据库在 Android 手机上建立一个学生成绩管理系统,演示记录的增、删、查、改等操作。

用户图形界面设计如图 9-2 所示。

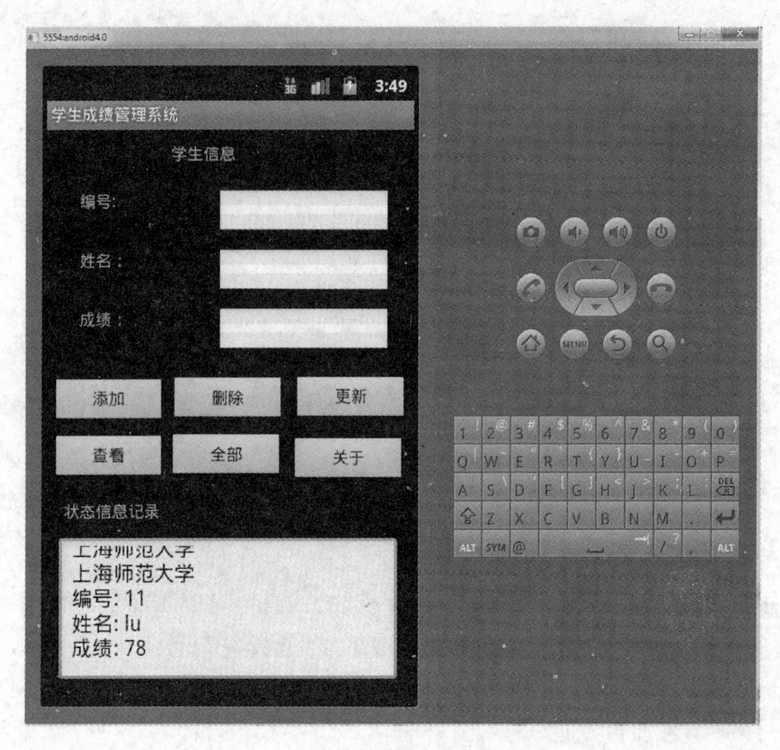

图 9-2　系统界面

Main. xml 内容如下：

```
<?xml version = "1.0" encoding = "utf - 8"?>
< AbsoluteLayout xmlns:android = "http://schemas.android.com/apk/res/android"
    android:id = "@ + id/myLayout"
    android:layout_width = "fill_parent"
    android:layout_height = "fill_parent"
    android:stretchColumns = "0" >

    < TextView
        android:layout_width = "wrap_content"
        android:layout_height = "wrap_content"
        android:layout_x = "110dp"
        android:layout_y = "10dp"
        android:text = "@string/title" />

    < TextView
        android:layout_width = "wrap_content"
        android:layout_height = "wrap_content"
        android:layout_x = "30dp"
        android:layout_y = "50dp"
        android:text = "@string/roll_no" />

    < EditText
        android:id = "@ + id/editRollno"
```

```
    android:layout_width = "150dp"
    android:layout_height = "40dp"
    android:layout_x = "150dp"
    android:layout_y = "50dp"
    android:inputType = "number" />

<TextView
    android:layout_width = "wrap_content"
    android:layout_height = "wrap_content"
    android:layout_x = "30dp"
    android:layout_y = "100dp"
    android:text = "@string/name" />

<EditText
    android:id = "@ + id/editName"
    android:layout_width = "150dp"
    android:layout_height = "40dp"
    android:layout_x = "150dp"
    android:layout_y = "100dp"
    android:inputType = "text" />

<TextView
    android:layout_width = "wrap_content"
    android:layout_height = "wrap_content"
    android:layout_x = "30dp"
    android:layout_y = "150dp"
    android:text = "@string/marks" />

<EditText
    android:id = "@ + id/editMarks"
    android:layout_width = "150dp"
    android:layout_height = "40dp"
    android:layout_x = "150dp"
    android:layout_y = "150dp"
    android:inputType = "number" />



<Button
    android:id = "@ + id/btnAdd"
    android:layout_width = "100dp"
    android:layout_height = "40dp"
    android:layout_x = "5dp"
    android:layout_y = "209dp"
    android:text = "@string/add" />

<Button
    android:id = "@ + id/btnModify"
```

```
        android:layout_width = "100dp"
        android:layout_height = "40dp"
        android:layout_x = "215dp"
        android:layout_y = "207dp"
        android:text = "@string/modify" />

    <Button
        android:id = "@ + id/btnDelete"
        android:layout_width = "100dp"
        android:layout_height = "40dp"
        android:layout_x = "109dp"
        android:layout_y = "208dp"
        android:text = "@string/delete" />

    <Button
        android:id = "@ + id/btnView"
        android:layout_width = "100dp"
        android:layout_height = "40dp"
        android:layout_x = "5dp"
        android:layout_y = "257dp"
        android:text = "@string/view" />

    <Button
        android:id = "@ + id/btnViewAll"
        android:layout_width = "100dp"
        android:layout_height = "40dp"
        android:layout_x = "108dp"
        android:layout_y = "256dp"
        android:text = "@string/view_all" />

    <Button
        android:id = "@ + id/btnShowInfo"
        android:layout_width = "100dp"
        android:layout_height = "40dp"
        android:layout_x = "213dp"
        android:layout_y = "259dp"
        android:text = "@string/show_info" />

    <EditText
        android:id = "@ + id/logMsg"
        android:layout_width = "294dp"
        android:layout_height = "124dp"
        android:layout_x = "12dp"
        android:layout_y = "344dp"
        android:ems = "10"
        android:inputType = "textMultiLine" >

        <requestFocus />
    </EditText>
```

```
    <TextView
        android:id="@+id/textView1"
        android:layout_width="wrap_content"
        android:layout_height="wrap_content"
        android:layout_x="16dp"
        android:layout_y="312dp"
        android:text="@string/logInfo" />

</AbsoluteLayout>
```

MyApp.java 内容如下：

```
public class MyApp extends Activity implements OnClickListener {
    EditText editRollno, editName, editMarks, logMsg;
    Button btnAdd, btnDelete, btnModify, btnView, btnViewAll, btnShowInfo;
    SQLiteDatabase db;

    /** Called when the activity is first created. */
    @Override
    public void onCreate(Bundle savedInstanceState) {
        super.onCreate(savedInstanceState);
        setContentView(R.layout.main);
        initUI();
        initDataBase();

    }

    public void initUI() {
        editRollno = (EditText) findViewById(R.id.editRollno);
        editName = (EditText) findViewById(R.id.editName);
        editMarks = (EditText) findViewById(R.id.editMarks);
        logMsg = (EditText) findViewById(R.id.logMsg);
        btnAdd = (Button) findViewById(R.id.btnAdd);
        btnDelete = (Button) findViewById(R.id.btnDelete);
        btnModify = (Button) findViewById(R.id.btnModify);
        btnView = (Button) findViewById(R.id.btnView);
        btnViewAll = (Button) findViewById(R.id.btnViewAll);
        btnShowInfo = (Button) findViewById(R.id.btnShowInfo);
        btnAdd.setOnClickListener(this);
        btnDelete.setOnClickListener(this);
        btnModify.setOnClickListener(this);
        btnView.setOnClickListener(this);
        btnViewAll.setOnClickListener(this);
        btnShowInfo.setOnClickListener(this);
    }


    public void onClick(View view) {

        switch (view.getId()) {
```

```
            case R.id.btnAdd:
                addRecord();
                break;
            case R.id.btnDelete:
                delRecord();
                break;
            case R.id.btnModify:
                modiRecord();
                break;
            case R.id.btnView:
                viewRecord();
                break;
            case R.id.btnViewAll:
                viewAllRecords();
                break;
            default:
                showMessage("学生成绩管理系统：", "上海师范大学");
                break;
            }

    }

//数据库的初始化:

public void initDataBase() {
        db = openOrCreateDatabase("StudentDB", Context.MODE_PRIVATE, null);
        db.execSQL("CREATE TABLE IF NOT EXISTS student(rollno VARCHAR, name VARCHAR, marks
VARCHAR);");
    }

//增加一条记录:

    public boolean addRecord() {
        if (editRollno.getText().toString().trim().length() == 0
                || editName.getText().toString().trim().length() == 0
                || editMarks.getText().toString().trim().length() == 0) {
            showMessage("Error", "请输入所有字段数据");
            return false;
        }
        db.execSQL("INSERT INTO student VALUES('" + editRollno.getText()
                + "','" + editName.getText() + "','" + editMarks.getText()
                + "');");
        showMessage("Success", "记录已经添加");
        clearText();
        return true;
    }

//删除一条记录:
```

```
    public boolean delRecord() {
        if (editRollno.getText().toString().trim().length() == 0) {
            showMessage("Error", "Please enter Rollno");
            return false;
        }
        Cursor c = db.rawQuery("SELECT * FROM student WHERE rollno = '"
                + editRollno.getText() + "'", null);
        if (c.moveToFirst()) {
            db.execSQL("DELETE FROM student WHERE rollno = '"
                    + editRollno.getText() + "'");
            showMessage("Success", "Record Deleted");
        } else {
            showMessage("Error", "查无此人的编号");
        }
        clearText();
        return true;
    }
```

//修改一条记录:

```
    public boolean modiRecord() {
        if (editRollno.getText().toString().trim().length() == 0) {
            showMessage("Error", "请输入学生编号");
            return false;
        }
        Cursor c = db.rawQuery("SELECT * FROM student WHERE rollno = '"
                + editRollno.getText() + "'", null);
        if (c.moveToFirst()) {
            db.execSQL("UPDATE student SET name = '" + editName.getText()
                    + "',marks = '" + editMarks.getText() + "' WHERE rollno = '"
                    + editRollno.getText() + "'");
            showMessage("Success", "记录已经修改成功");
        } else {
            showMessage("Error", "查无此人的编号");
        }
        clearText();
        return true;
    }
```

//查找显示一条记录:

```
    public boolean viewRecord() {
        if (editRollno.getText().toString().trim().length() == 0) {
            viewAllRecords();
            return false;
        }
        Cursor c = db.rawQuery("SELECT * FROM student WHERE rollno = '"
                + editRollno.getText() + "'", null);
        if (c.moveToFirst()) {
```

```
                editName.setText(c.getString(1));
                editMarks.setText(c.getString(2));
            } else {
                showMessage("Error", "查无此人的编号");
                clearText();
            }
            return true;
        }

    //显示全部记录:

        public boolean viewAllRecords() {
            Cursor c = db.rawQuery("SELECT * FROM student", null);
            if (c.getCount() == 0) {
                showMessage("Error", "数据库无记录");
                return false;
            }
            StringBuffer buffer = new StringBuffer();
            while (c.moveToNext()) {
                buffer.append("编号: " + c.getString(0) + "\n");
                buffer.append("姓名: " + c.getString(1) + "\n");
                buffer.append("成绩: " + c.getString(2) + "\n\n");
            }
            showMessage("学生信息: ", buffer.toString());
            return true;
        }

    //显示对话框
        public void showMessage(String title, String message) {
            Builder builder = new Builder(this);
            builder.setCancelable(true);
            builder.setTitle(title);
            builder.setMessage(message);
            logMsg.append(message + "\n");
            builder.show();
        }
    //清空 UI 文本输入框内容:

        public void clearText() {
            editRollno.setText("");
            editName.setText("");
            editMarks.setText("");
            editRollno.requestFocus();
        }
    }
```

系统运行结果如图 9-3 所示。

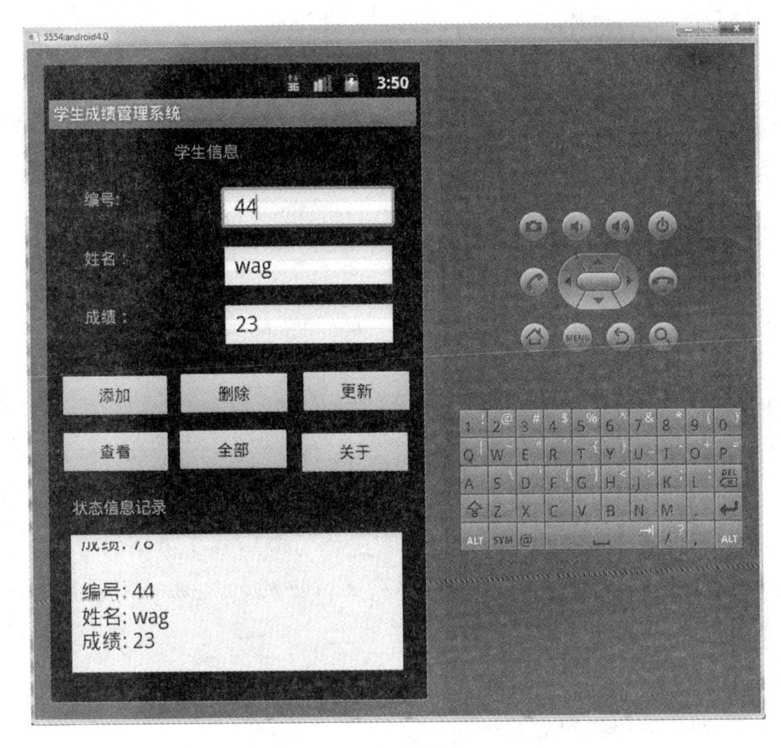

图 9-3　系统运行界面

9.7　本章小结

通过本章的学习，我们已经掌握了 Android 应用程序中文件与数据库的安全模型、文件与数据库的创建以及读写方式。

9.8　习题与课外阅读

9.8.1　习题

(1) 简述 Android 文件系统安全模型。

(2) 简述 SQLite 数据库与现有的网络数据库(如 Mysql)安全权限区别。

(3) 创建一个文件，完成将“hello world”字符串写入文件中。

(4) 试创建一个 SQLite 数据库，并实现数据记录的增删改查等功能。

(5) 简述 ADB 协议原理及功能。

9.8.2　课外阅读

(1) 访问下列技术网站，了解一下 SQLite 数据相关信息：

http://www.sqlite.org

(2) 访问下列技术网站，学习使用可视化 SQLite 管理工具：

http://www.sqliteexpert.com/

第10章 ContentProvider

ContentProvider(内容提供者)是 Android 中的四大组件之一,主要用于对外共享数据。在 Android 系统中,没有一个公共的内存区域,供多个应用共享存储数据,Android 中的 ContentProvider 机制可支持在多个应用中存储和读取数据,就是通过 ContentProvider 把应用中的数据共享给其他应用访问,这也是跨进程与应用共享数据的唯一方式,其他应用可以通过 ContentProvider 对指定应用中的数据进行操作。Android 系统已经预置了几种 ContentProvider,提供了途径。

本章学习目标:

- 掌握 ContentProvider 概念;
- 掌握读写系统提供 ContentProvider 的方法。

10.1 ContentProvider 简介

Android 系统为了保证安全,没有提供一个公共的内存区域,供多个应用共享存储数据。而是提供了 ContentProvider 机制来支持在多个应用之间共享存储和读取数据,来实现在不同应用间共享数据。ContentProvider 的主要功能如下:

- 访问现有的资源。开发者可以利用 Android 系统提供的 ContentProvider,访问系统提供的数据,比如音频、视频、图片和私人通讯录等,当然前提是已获得 ContentProvider 适当的读取权限。Android 系统提供的 ContentProvider 可以在 android.provider 包里面找到。
- 共享现有的资源。开发者可以为自己的应用编写 ContentProvider,让其他应用来访问开发者应用程序中的数据。如果想通过 ContentProvider 方式把自己的数据共享给其他应用,有两种办法:

  (1) 创建自己的 ContentProvider,需要继承 ContentProvider 类,实现数据添加(insert)、删除(delete)、查询(query)、修改(update)等方法;

  (2) 利用现有的 ContentProvider。如果数据和 Android 系统提供的现有 ContentProvider 数据结构一致,可以将数据写到已存在的 ContentProvider 中。该应用程序必须具有读写该 ContentProvider 的权限。

• ContentProvider 展示数据类似一个单个数据库表。其中：

(1) 每行有个带唯一值的数字字段，名为_ID，可用于对表中指定记录的定位；

(2) ContentProvider 返回的数据结构，是 Cursor 对象，有点类似 JDBC 的 ResultSet。

10.2 ContentResolver 简介

所有的 ContentProvider 都需要实现相同的接口用于查询 ContentProvider 并返回数据，也包括添加(insert)、删除(delete)、查询(query)、修改(update)数据。ContentProvider 的用户都不能直接访问到 ContentProvider 实例，只能通过 ContentResolver 在中间代理来实现对 ContentProvider 的操作(见图 10-1)。

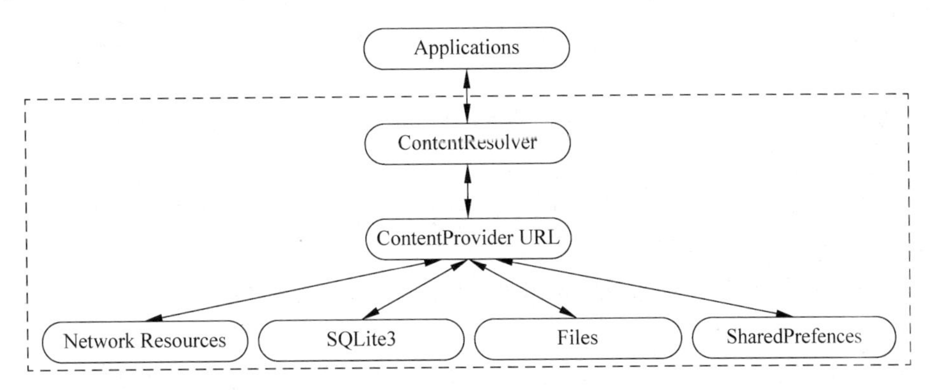

图 10-1 应用与 ContentResolver 和 ContentProvider 及数据文件之间的关系

ContentProvider 是单例模式的，当多个应用程序通过 ContentResolver 操作 ContentProvider 的数据时，ContentResolver 调用的数据将会委托给同一个 ContentProvider 处理。

应用程序通过 Activity 的 getContentResovler()成员方法获得一个 ContentResolver 的实例：

```
ContentResolver cr = getContentResolver();
```

ContentResolver 是通过 URI 来查询 ContentProvider 中提供的数据。ContentResolver 的主要方法，如表 10-1 所示。

表 10-1 ContentResolver 方法

方法名称	作用
final Uri insert (Uri url, ContentValues values)	插入一行数据到给定的 url 数据表中
final int delete (Uri url, String where, String [ ] selectionArgs)	删除 url 数据表中特定数据行
final Cursor query (Uri uri, String[] projection, String selection, String[] selectionArgs, String sortOrder)	查询给定的 url 数据，返回 Cursor 数据集合
final int update (Uri uri, ContentValues values, String where, String[] selectionArgs)	更新给定 url 数据表中行数据

10.3 ContentProvider 数据的 URI 表达

每个 ContentProvider 定义一个唯一的公开的 URI，指向它的数据集。一个 ContentProvider 可以包含多个数据集(可以看作多张表)，这样，就需要有多个 URI 与每个数据集对应。即：URI 代表了要操作的数据，URI 主要包含了两部分信息(见图 10-2)：

第一，被操作的 ContentProvider 对象；

第二，被操作的 ContentProvider 对象数据。

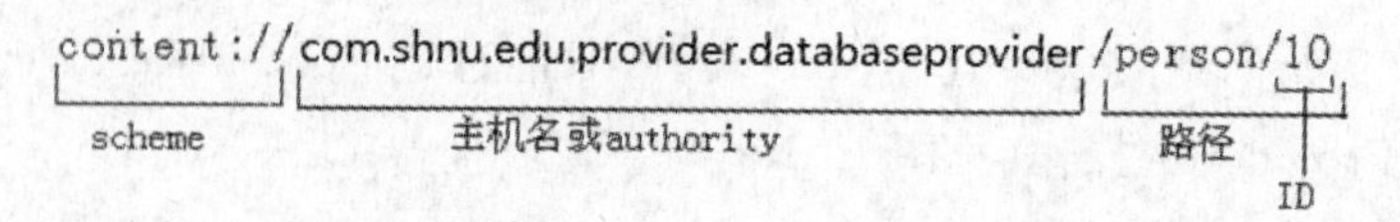

图 10-2　ContentProvider 数据示例

(1) scheme：ContentProvider(内容提供者)标准前缀，已经由 Android 所规定为：content://，无法改变的。

(2) 主机名(或 Authority)：用于唯一标识这个 ContentProvider，外部调用者可以根据这个标识来找到它。

(3) 路径(path)：可以用来表示我们要操作的数据，路径的构建应根据业务而定，具体如下：

- 要操作 person 表中 id 为 10 的记录，可以构建这样的路径：/ person/10。
- 要操作 person 表中 id 为 10 的记录的 name 字段，可以构建这样的路径：/person/10/name。
- 要操作 person 表中的所有记录，可以构建这样的路径：/ person。
- 要操作的数据不一定来自数据库，如要操作 xml 文件中 person 节点下的 name 节点，可以构建这样的路径：/person/name。

通常情况下，URI 是按照字符串的形式表达的，字符串需要转化成 URI 对象。可以使用 Uri 类中的 parse()方法，如下：

```
Uri uri = Uri.parse("content://com.shnu.edu.provider.databaseprovider/person")
```

另外，Android 提供了更为方便的方法，让开发者不需要自己拼接上面这样的 URI 字符串，如：

```
Uri myPerson = ContentUris.withAppendedId(People.CONTENT_URI, 23);
```

或者：

```
Uri myPerson = Uri.withAppendedPath(People.CONTENT_URI, "23");
```

二者的区别是一个接收整数类型的 ID 值，一个接收字符串类型。

定义一个 ContentProvider，最好使用常量。Android 定义了 CONTENT_URI 常量用于 URI(见表 10-2)。

表 10-2　Android 系统管理联系人的 URI

URI 名称	作　　用
ContactsContract. Contacts. CONTENT_URI	管理联系人的 Uri
ContactsContract. CommonDataKinds. Phone. CONTENT_URI	管理联系人的电话的 Uri
ContactsContract. CommonDataKinds. Email. CONTENT_URI	管理联系人的 Email 的 Uri

注：Contacts 有两个表，分别是 rawContact 和 Data。

rawContact 记录了用户的 id 和 name。

id 为 ContactsContract. Contacts. _ID。

name 为 ContactContract. Contracts. DISPLAY_NAME。

电话信息表的外键 id 为 ContactsContract. CommonDataKinds. Phone. CONTACT_ID。

电话号码栏名称为 ContactsContract. CommonDataKinds. Phone. NUMBER。

data 表中 E-mail 地址栏名称为 ContactsContract. CommonDataKinds. Email. DATA。

其外键栏为 ContactsContract. CommonDataKinds. Email. CONTACT_ID。

Android 系统对多媒体提供了许多 URI，见表 10-3。

表 10-3　Android 为多媒体提供的 ContentProvider 的 URI

函 数 名 称	作　　用
MediaStore. Audio. Media. EXTERNAL_CONTENT_URI	存储在 SD 卡上的音频文件
MediaStore. Audio. Media. INTERNAL_CONTENT_URI	存储在手机内部存储器上的音频文件
MediaStore. Audio. Images. EXTERNAL_CONTENT_URI	SD 卡上的图片文件内容
MediaStore. Audio. Images. INTERNAL_CONTENT_URI	手机内部存储器上的图片
MediaStore. Audio. Video. EXTERNAL_CONTENT_URI	SD 卡上的视频
MediaStore. Audio. Video. INTERNAL_CONTENT_URI	手机内部存储器上的视频

10.4　利用 ContentProvider 显示通讯录数据

如果要使用一个 ContentProvider 查询，需要以下信息：

(1) 提供 ContentProvider 对应的 URI，其中 URI 是必需的，其他是可选的，如果系统能找到 URI 对应的 ContentProvider 将返回一个 Cursor 对象。如果需要查询 ContentProvider 数据集的特定记录(行)，还需要提供该记录的 ID 的值。

(2) 返回结果的字段名称和这些字段的数据类型。

可以使用两种查询方法：

(1) Cursor c = getContentResolver(). query ()方法。

(2) Cursor c = Activity. managedQuery()方法。

两者的方法参数完全一样，查询过程和返回值也是相同的。

区别是：通过 Activity. managedQuery()方法，不但获取到 Cursor 对象，而且能够管理 Cursor 对象的生命周期。比如，当 Activity 暂停(pause)的时候，可以卸载该 Cursor 对象，当 Activity restart 的时候重新查询。另外，也可以对一个没有处于 Activity 管理的 Cursor

对象做成被 Activity 管理的，通过调用 Activity. startManaginCursor()方法。类似这样：

```
public final Cursor managedQuery(Uri uri,
                        String[] projection,
                        String selection,
                        String[] selectionArgs,
                        String sortOrder)
    { Cursor c = getContentResolver().query(uri, projection, selection, selectionArgs,
sortOrder);
     if (c != null) {
                startManagingCursor(c);  }
    return c;
    }
```

getContentResolver(). query 相关参数见表 10-4。

表 10-4 getContentResolver(). query(uri, projection, selection, selectionArgs, sortOrder)参数说明

参数名称	说　明
uri	用于 Content Provider 查询的 URI，也就是说从这个 URI 中获取数据。 例如，Uri uri = Contacts. People. CONTENT_URI； //联系人列表 URI
projection	用于标识 uri 中有哪些 columns 需要包含在返回的 Cursor 对象中。 例如，String [ ] projection = { Contacts. PeopleColumns. NAME, Contacts. PeopleColumns. NOTES }；
selection	作为查询的过滤参数(过滤出符合 selection 的数据)，类似 SQL 中 Where 语句之后的条件选择。 例如，String selection = Contacts. People. NAME + "=?" //查询条件
selectionArgs	查询条件参数，配合 selection 参数使用。 例如，String[] selectionArgs = {"James", "Jack"}；//查询条件参数
sortOrder	查询结果的排序方式(按查询列(projection 参数中的 columns)中的某个 column)排序)。 例如，String sortOrder = Contacts. PeopleColumns. NAME；//查询结果按指定的名称排序

返回值是一个包含指定数据的 Cursor 对象。Cursor 和 JDBC 的 ResultSet 对象类似，需要操作游标遍历结果集，在每行再通过列名获取到列的值，可以通过 getString()、getInt()、getFloat()等方法获取值。如：

```
private Uri contactsURI = ContactsContract.Contacts.CONTENT_URI;
private Uri rawURI = ContactsContract.RawContacts.CONTENT_URI;
private Uri dataURI = ContactsContract.Data.CONTENT_URI;
private Uri phoneURI = ContactsContract.CommonDataKinds.Phone.CONTENT_URI;
private Uri emailURI = ContactsContract.CommonDataKinds.Email.CONTENT_URI;
```

显示通讯录中联系人的姓名。

```
public void readContacts() {
        Cursor cursor = cr.query(contactsURI, null, null, null, null);
        /* 操作游标,获取数据 */
```

```
        for (cursor.moveToFirst(); !cursor.isAfterLast(); cursor.moveToNext()) {
                StringBuffer sb = new StringBuffer();
                /* 获取联系人 ID */
                 String contactID = cursor.getString(cursor.getColumnIndex(ContactsContract.
Contacts._ID));
                sb.append("ID:" + contactID + "\n");
                /* 获取联系人姓名 */
                 String name = cursor.getString(cursor.getColumnIndex(ContactsContract.
Contacts.DISPLAY_NAME));sb.append("姓名: " + name + "\n");
                tv.append(sb.toString());
        }
        cursor.close();
    }
```

10.5 利用 ContentProvider 添加通讯录数据

```
public void insertContacts(String name, String phoneMobile,
        String phoneWorke, String email) {

    /*
     * 首先: 需要向 RawContacts.CONTENT_URI
     * 执行一个空值的插入,目的是获取系统给这条记录自动设定的 rawConatactId,
     * 这是后面插入 data 表的依据,只有执行空值的插入,才能使插入的联系人在通讯录里可见
     */
    ContentValues values = new ContentValues();
    /* 向 RawContacts.CONTENT_URI 执行一个空值的插入,返回 rawContactId */
    Uri rawContactUri = cr.insert(rawURI, values);
    /* 从 Uri 路径中获取 ID 的部分 */
    long rawContactId = ContentUris.parseId(rawContactUri);

    values.clear();

    /* data 表中的数据结构特点: 每个数据信息以行进行保存,所以每次添加一行数据 */

    /* 向 data 表中插入姓名 */
    values.put(Data.RAW_CONTACT_ID, rawContactId); // ID
    values.put(Data.MIMETYPE, StructuredName.CONTENT_ITEM_TYPE); // 内容的类型
    values.put(StructuredName.GIVEN_NAME, name);
    cr.insert(dataURI, values);
    values.clear();

    /* 向 data 表中插入移动电话 */
    values.put(Data.RAW_CONTACT_ID, rawContactId);
    values.put(Data.MIMETYPE, Phone.CONTENT_ITEM_TYPE);
    values.put(Phone.NUMBER, phoneMobile);
    values.put(Phone.TYPE, Phone.TYPE_MOBILE); // 电话的类型: 工作电话 移动电话 家庭电话
    cr.insert(dataURI, values);
```

```
        values.clear();

        /* 向 data 表中插入工作电话 */
        values.put(Data.RAW_CONTACT_ID, rawContactId);
        values.put(Data.MIMETYPE, Phone.CONTENT_ITEM_TYPE);
        values.put(Phone.NUMBER, phoneWorke);
        values.put(Phone.TYPE, Phone.TYPE_WORK); // 电话的类型: 工作电话 移动电话 家庭电话
        cr.insert(dataURI, values);
        values.clear();

        /* 向 data 表中添加 Email */
        values.put(Data.RAW_CONTACT_ID, rawContactId);
        values.put(Data.MIMETYPE, Email.CONTENT_ITEM_TYPE);
        values.put(Email.DATA, email);
        values.put(Email.TYPE, Email.TYPE_WORK); // 电话的类型: 工作电话 移动电话 家庭电话
        this.getContentResolver().insert(
                    android.provider.ContactsContract.Data.CONTENT_URI, values);
        values.clear();
        String s = name + "\n Phone:" + phoneMobile + " " + phoneWorke
                    + "\n Email:" + email;
        Toast.makeText(MainActivity.this, s + "\n 数据添加成功!", 1000).show();

    }
```

10.6 利用 ContentProvider 删除通讯录数据

```
public void deleteContacts(String name) {

        /* 思路: 删除一个联系人的所有,则根据 RAW_CONTACT_ID 进行删除 */
        String whereClause = ContactsContract.Data.DISPLAY_NAME + " = ?";
        String[] whereArgs = { name };
        cr.delete(rawURI, whereClause, whereArgs);
        Toast.makeText(MainActivity.this, name + " 数据删除成功", 1000).show();
  }
ContentValues values = new ContentValues();
values.put(People.NAME, "Abraham Lincoln");
Uri uri = getContentResolver().insert(People.CONTENT_URI, values);
```

10.7 利用 ContentProvider 更新通讯录数据

```
public void updateContacts(String name, String newNumber) {

        /* 获取 ID,需要修改的联系人 ID,然后确定修改信息 */
String where = ContactsContract.Data.DISPLAY_NAME + " = ?";
String[] whereArgs = { name };
```

```
        ContentValues values = new ContentValues();
        values.put(ContactsContract.CommonDataKinds.Phone.DATA, newNumber);
        cr.update(dataURI, values, where, whereArgs);
        Toast.makeText(MainActivity.this, "修改成功", 1000).show();
        readContacts();

    }

}
```

【例 10-1】 MainActivity。

```
package com.example.contactdemo;

import android.app.Activity;
import android.content.ContentResolver;
import android.content.ContentUris;
import android.content.ContentValues;
import android.database.Cursor;
import android.net.Uri;
import android.os.Bundle;

import android.provider.ContactsContract;
import android.provider.ContactsContract.CommonDataKinds;
import android.provider.ContactsContract.CommonDataKinds.Email;
import android.provider.ContactsContract.CommonDataKinds.Phone;
import android.provider.ContactsContract.CommonDataKinds.StructuredName;
import android.provider.ContactsContract.Data;
import android.provider.ContactsContract.RawContacts;
import android.view.View;
import android.view.View.OnClickListener;
import android.widget.Button;
import android.widget.EditText;
import android.widget.Toast;

public class MainActivity extends Activity {

    private Button insert, delete, update, read;
    private EditText tv;

    /* 获取 ContentResolver 对象,使用 getContentResolver()方法 */
    private ContentResolver cr;
    private Uri contactsURI = ContactsContract.Contacts.CONTENT_URI;
    private Uri rawURI = ContactsContract.RawContacts.CONTENT_URI;
    private Uri dataURI = ContactsContract.Data.CONTENT_URI;
    private Uri phoneURI = ContactsContract.CommonDataKinds.Phone.CONTENT_URI;
    private Uri emailURI = ContactsContract.CommonDataKinds.Email.CONTENT_URI;

    protected void init() {
        tv = (EditText) this.findViewById(R.id.editText1);
```

```
        insert = (Button) findViewById(R.id.insertContact);
        delete = (Button) findViewById(R.id.deleteContacts);
        update = (Button) findViewById(R.id.updateContacts);
        read = (Button) findViewById(R.id.readContact);
        cr = this.getContentResolver();
    }

    protected void onCreate(Bundle savedInstanceState) {
        super.onCreate(savedInstanceState);
        setContentView(R.layout.activity_main);
        /* 初始化组件对象 */
        init();

        /* 为 read 设置按钮点击事件监听器 */
        read.setOnClickListener(new OnClickListener() {

            public void onClick(View v) {
                readContacts();
            }
        });

        insert.setOnClickListener(new OnClickListener() {

            @Override
            public void onClick(View v) {
                // TODO Auto-generated method stub
                insertContacts("James", "18988888888", "02164328888",
                                "8888@qq.com");
            }
        });

        delete.setOnClickListener(new OnClickListener() {

            @Override
            public void onClick(View v) {
                // TODO Auto-generated method stub
                deleteContacts("James");
            }
        });

        update.setOnClickListener(new OnClickListener() {

            @Override
            public void onClick(View v) {
                // TODO Auto-generated method stub
                updateContacts("James", "1234567");
            }
        });
    }
```

```
/* 读取联系人信息,姓名,电话,Email */
public void readContacts() {

    Cursor cursor = cr.query(contactsURI, null, null, null, null);

    /* 操作游标,获取数据 */
    for (cursor.moveToFirst(); !cursor.isAfterLast(); cursor.moveToNext()) {

        StringBuffer sb = new StringBuffer();
        /* 获取联系人 ID */
        String contactID = cursor.getString(cursor
                .getColumnIndex(ContactsContract.Contacts._ID));
        sb.append("ID:" + contactID + "\n");
        /* 获取联系人姓名 */

        String name = cursor.getString(cursor
                .getColumnIndex(ContactsContract.Contacts.DISPLAY_NAME));
        sb.append("姓名: " + name + "\n");
        /* 利用联系人 ID 获取电话号码 */
        sb.append(getPhone(contactID));
        /* 利用联系人 ID 获取 Email */
        sb.append(getEmail(contactID));
        tv.setText("");
        tv.append(sb.toString());

    }
    cursor.close();

}

// 从 ContactsContract.CommonDataKinds.Phone.CONTENT_URI 中获取电话号码
// 在 ContactsContract.CommonDataKinds.Phone.CONTACT_ID 获取与 contactID 对应的电话号码
public String getPhone(String contactID) {
    StringBuffer sb = new StringBuffer();
    Cursor phone = cr.query(phoneURI, null,
            ContactsContract.CommonDataKinds.Phone.CONTACT_ID + " = ?",
            new String[] { contactID }, null);
    sb.append("Phone:");
    while (phone.moveToNext()) {
        String phoneNumber = phone
                .getString(phone
.getColumnIndex(ContactsContract.CommonDataKinds.Phone.NUMBER));
        sb.append("\t" + phoneNumber + "\n");

    }
    /* 游标使用后要关闭 */
    phone.close();
    return sb.toString();
}
```

```
    // 从 ContactsContract.CommonDataKinds.Email.CONTENT_URI 中获取 Email
    // ContactsContract.CommonDataKinds.Email.CONTACT_ID 获取与 contactID 对应 Email;
    public String getEmail(String contactID) {
        StringBuffer sb = new StringBuffer();
        Cursor email = cr.query(emailURI, null,
                    ContactsContract.CommonDataKinds.Email.CONTACT_ID + " = ?",
                    new String[] { contactID }, null);
        sb.append("Email:");
        while (email.moveToNext()) {
            String emailAddress = email
                        .getString(email
.getColumnIndex(ContactsContract.CommonDataKinds.Email.DATA));
            sb.append("\t" + emailAddress + "\n");

        }
        email.close();
        return sb.toString();
    }

    /* 添加联系人 */
    public void insertContacts(String name, String phoneMobile,
            String phoneWorke, String email) {

        /*
         * 首先：需要向 RawContacts.CONTENT_URI
         * 执行一个空值的插入,目的是获取系统给这条记录自动设定的 rawConatactId,
         * 这是后面插入 data 表的依据,只有执行空值的插入,才能使插入的联系人在通讯录里可见
         */
        ContentValues values = new ContentValues();

        /* 向 RawContacts.CONTENT_URI 执行一个空值的插入,返回 rawContactId */
        Uri rawContactUri = cr.insert(rawURI, values);

        /* 从 Uri 路径中获取 ID 的部分 */
        long rawContactId = ContentUris.parseId(rawContactUri);

        values.clear();

        /* data 表中的数据结构特点：每个数据信息以行进行保存,所以每次添加一行数据 */

        /* 向 data 表中插入姓名 */
        values.put(Data.RAW_CONTACT_ID, rawContactId); // ID
        values.put(Data.MIMETYPE, StructuredName.CONTENT_ITEM_TYPE); // 内容的类型
        values.put(StructuredName.GIVEN_NAME, name);
        cr.insert(dataURI, values);
        values.clear();

        /* 向 data 表中插入移动电话 */
        values.put(Data.RAW_CONTACT_ID, rawContactId);
```

```
        values.put(Data.MIMETYPE, Phone.CONTENT_ITEM_TYPE);
        values.put(Phone.NUMBER, phoneMobile);
        values.put(Phone.TYPE, Phone.TYPE_MOBILE); // 电话的类型：工作电话 移动电话 家庭电话
        cr.insert(dataURI, values);
        values.clear();

        /* 向data表中插入工作电话 */
        values.put(Data.RAW_CONTACT_ID, rawContactId);
        values.put(Data.MIMETYPE, Phone.CONTENT_ITEM_TYPE);
        values.put(Phone.NUMBER, phoneWorke);
        values.put(Phone.TYPE, Phone.TYPE_WORK); // 电话的类型：工作电话 移动电话 家庭电话
        cr.insert(dataURI, values);
        values.clear();

        /* 向data表中添加Email */
        values.put(Data.RAW_CONTACT_ID, rawContactId);
        values.put(Data.MIMETYPE, Email.CONTENT_ITEM_TYPE);
        values.put(Email.DATA, email);
        values.put(Email.TYPE, Email.TYPE_WORK); // 电话的类型：工作电话 移动电话 家庭电话
        this.getContentResolver().insert(
                        android.provider.ContactsContract.Data.CONTENT_URI, values);
        values.clear();
        String s = name + "\n Phone:" + phoneMobile + " " + phoneWorke
                        + "\n Email:" + email;
        Toast.makeText(MainActivity.this, s + "\n 数据添加成功!", 1000).show();

    }

    /* 删除 */
    public void deleteContacts(String name) {

        /* 思路：删除一个联系人的所有,则根据RAW_CONTACT_ID进行删除 */
        String whereClause = ContactsContract.Data.DISPLAY_NAME + " = ?";
        String[] whereArgs = { name };
        cr.delete(rawURI, whereClause, whereArgs);
        Toast.makeText(MainActivity.this, name + " 数据删除成功", 1000).show();
    }

    /* 修改 */
    public void updateContacts(String name, String newNumber) {

        /* 获取ID,需要修改的联系人ID,然后确定修改信息 */
//      ContentValues values = new ContentValues();
//      values.put(Phone.NUMBER, newNumber);
    String where = ContactsContract.Data.DISPLAY_NAME + " = ?";
    String[] whereArgs = { name };
//      cr.update(dataURI, values, whereClause, whereArgs);

//      String where = ContactsContract.Data.DISPLAY_NAME + " = ?AND " +
//                      ContactsContract.Data.MIMETYPE + " = ?AND " +
```

```
//                          CommonDataKinds.Phone.TYPE_MOBILE + " = ?";
//          String [ ]selectionArgs = new String[ ]{name,
//                  CommonDataKinds.Phone.CONTENT_ITEM_TYPE,
//                  "" + CommonDataKinds.Phone.TYPE_MOBILE};
        ContentValues values =  new ContentValues();
        values.put(ContactsContract.CommonDataKinds.Phone.DATA, newNumber);
        cr.update(dataURI, values, where, whereArgs);
        Toast.makeText(MainActivity.this, "修改成功", 1000).show();
        readContacts();

    }

}
```

activity_main.xml：

```
<?xml version = "1.0" encoding = "utf - 8"?>
< RelativeLayout xmlns:android = "http://schemas.android.com/apk/res/android"
    android:id = "@ + id/RelativeLayout1"
    android:layout_width = "match_parent"
    android:layout_height = "match_parent"
    android:orientation = "vertical" >

    < Button
        android:id = "@ + id/updateContacts"
        android:layout_width = "wrap_content"
        android:layout_height = "wrap_content"
        android:layout_alignBaseline = "@ + id/deleteContacts"
        android:layout_alignBottom = "@ + id/deleteContacts"
        android:layout_toRightOf = "@ + id/deleteContacts"
        android:text = "更新" />

    < EditText
        android:id = "@ + id/editText1"
        android:layout_width = "match_parent"
        android:layout_height = "wrap_content"
        android:layout_alignParentLeft = "true"
        android:layout_alignParentRight = "true"
        android:layout_below = "@ + id/insertContact"
        android:layout_marginTop = "93dp"
        android:ems = "10"
        android:inputType = "textMultiLine" />

    < Button
        android:id = "@ + id/insertContact"
        android:layout_width = "wrap_content"
        android:layout_height = "wrap_content"
        android:layout_alignBaseline = "@ + id/readContact"
        android:layout_alignBottom = "@ + id/readContact"
        android:layout_toLeftOf = "@ + id/readContact"
```

```
        android:text = "插入" />

    <Button
        android:id = "@ + id/deleteContacts"
        android:layout_width = "wrap_content"
        android:layout_height = "wrap_content"
        android:layout_alignParentTop = "true"
        android:layout_centerHorizontal = "true"
        android:layout_marginTop = "134dp"
        android:text = "删除" />

    <Button
        android:id = "@ + id/readContact"
        android:layout_width = "wrap_content"
        android:layout_height = "wrap_content"
        android:layout_alignBaseline = "@ + id/deleteContacts"
        android:layout_alignBottom = "@ + id/deleteContacts"
        android:layout_toLeftOf = "@ + id/deleteContacts"
        android:text = "显示" />

</RelativeLayout>
```

AndroidManifest. xml：

```
  <?xml version = "1.0" encoding = "utf - 8"?>
<manifest xmlns:android = "http://schemas.android.com/apk/res/android"
    package = "com.example.contactdemo"
    android:versionCode = "1"
    android:versionName = "1.0" >

    <uses - sdk
        android:minSdkVersion = "8"
        android:targetSdkVersion = "17" />
    <uses - permission android:name = "android.permission.READ_CONTACTS"/>
    <uses - permission android:name = "android.permission.WRITE_CONTACTS"/>

    <application
        android:allowBackup = "true"
        android:icon = "@drawable/ic_launcher"
        android:label = "@string/app_name"
        android:theme = "@style/AppTheme" >
        <activity
            android:name = "com.example.contactdemo.MainActivity"
            android:label = "@string/app_name" >
            <intent - filter>
                <action android:name = "android.intent.action.MAIN" />

                <category android:name = "android.intent.category.LAUNCHER" />
            </intent - filter>
        </activity>
```

```
    </application>

</manifest>
```


10.8 本章小结

通过本章的学习，我们已经掌握了 Android 应用程序中通过 ContentProvider，来对通讯录进行实现数据添加(insert)、删除(delete)、查询(query)、修改(update)等方法。

10.9 习题与课外阅读

10.9.1 习题

(1) 简述 ContentProvider 与 ContentResolver 之间的关系。
(2) 编写一个程序，来统计手机用户每个月的通话时间。

10.9.2 课外阅读

访问下列技术网站，了解一下 Android 提供的其他 ContentProvider：
http://developer.android.com/training/index.html

第11章 Android 传感器

Android 手机应用的用户体验性至关重要，除了 UI 与用户交互外，Android 手机中内置了许多传感器，如加速度传感器、陀螺传感器、光线传感器等，这些传感器可以感知手机的位置、环境和物理状态等数据。利用这些数据，开发者可以做出交互性更好、体验性更好、更加智能化的 Android 应用程序。本章重点介绍 Android 传感器数据的采集，以及传感器相关的应用程序编写。

本章学习目标：

- 了解 Android 系统中的传感器类型；
- 掌握 SensorManager、Sensor 对象信息的获取方法；
- 掌握编写传感器数据采集程序的方法。

11.1 Android 系统中传感器介绍

Android 手机内置了许多传感器(Sensor)，自从 Android 系统 API Level 14 版本以后，可以支持多达二十种传感器，见表 11-1。

表 11-1 Android 系统支持的传感器种类

传感器类型常量(int)	数值	中 文 名 称
TYPE_ACCELEROMETER	1	加速度传感器
TYPE_MAGNETIC_FIELD	2	磁场传感器
TYPE_ORIENTATION	3	方向传感器。已过时。API level 8 后．用函数替代：SensorManager. getOrientation()
TYPE_GYROSCOPE	4	陀螺传感器
TYPE_LIGHT	5	光传感器
TYPE_PRESSURE	6	压力传感器
TYPE_TEMPERATURE	7	温度传感器。已过时。API level 14 后用 Sensor. TYPE_AMBIENT_TEMPERATURE
TYPE_PROXIMITY	8	距离传感器
TYPE_GRAVITY	9	重力传感器
TYPE_LINEAR_ACCELERATION	10	线性加速度传感器
TYPE_ROTATION_VECTOR	11	旋转矢量传感器
TYPE_RELATIVE_HUMIDITY	12	相对湿度传感器
TYPE_AMBIENT_TEMPERATURE	13	环境温度传感器
TYPE_MAGNETIC_FIELD_UNCALIBRATED	14	未校正磁场传感器

续表

传感器类型常量(int)	数值	中文名称
TYPE_GAME_ROTATION_VECTOR	15	未校正旋转矢量传感器
TYPE_GYROSCOPE_UNCALIBRATED	16	未校正陀螺传感器
TYPE_SIGNIFICANT_MOTION	17	运动触发传感器
TYPE_STEP_DETECTOR	18	步伐传感器
TYPE_STEP_COUNTER	19	计步传感器
TYPE_GEOMAGNETIC_ROTATION_VECTOR	20	地磁场旋转矢量传感器
TYPE_ALL	−1	所有传感器

对于具体特定 Android 设备支持哪些类型传感器，请参照相关厂商的硬件平台介绍。目前绝大多数 Android 手机都内置了加速度传感器(Accelerometer)、陀螺仪(Gyroscope)、环境光照传感器(Light)、磁力传感器(Magnetic field)、方向传感器(Orientation)、压力传感器(Pressure)、距离传感器(Proximity)和温度传感器(Temperature)等传感器。

开发者利用这些传感器可以动态感知 Android 手机的所在位置、状态、方向、加速表、磁场、距离、温度、所处的环境亮度光线等，从而开发出用户体验性更好 Android 应用。如 GPS 导航与位置服务应用、微信的“摇一摇”搜索功能，把 Android 应用做到“极简、极致”。Android 应用与传感器结合可以让 Android 智能手机的功能更加丰富多彩。

11.2 Android 系统中传感器信息的获取

Android 系统开发传感器相关程序主要用到 5 类或接口，见表 11-2。

表 11-2　Android 系统传感器相关的类

主要类	说明
SensorManager.java	实现传感器系统核心的管理类 SensorManager
Sensor.java	单一传感器的描述性文件 Sensor
SensorEvent.java	表示传感器系统的事件类 SensorEvent
SensorEventListener.java	传感器事件的监听者 SensorEventListener 接口
SensorListener.java	传感器的监听者 SensorListener 接口(不推荐使用)

在应用程序中，要使用特定传感器，首先获取传感器管理器对象 SensorManager，然后根据传感器的类型(见表 11-1)，获取传感器对象，获取传感器对象以后就可以使用传感器了。具体步骤如下：

(1) 传感器管理对象 SensorManager 的获得。

对传感器的访问之前，必须首先获取一个 SensorManager 对象。通常可以使用 Context.getSystemService()方法来获得一个 SensorManager 对象。

```
//从系统服务中获得传感器管理器
SensorManager mSensorManager = (SensorManager)getSystemService(SENSOR_SERVICE);
```

(2) Android 系统中所有传感器对象的获得。

取得 SensorManager 对象之后，可以通过 getSensorList()方法来获得需要的传感器类型，并将之保存到一个传感器列表中。

```
List<Sensor> allSensors = sensorManager.getSensorList(Sensor.TYPE_ALL);
```

(3) 传感器相关性能参数信息获得。

获取了传感器对象后，就可以使用如表 11-3 所示的方法获取传感器的相关信息。

表 11-3 传感器相关性能参数信息获取方法

传感器类	方 法	功能描述
Sensor	getName()	设备名称
	getPower()	功率
	getResolution()	精度
	getType()	传感器类型
	getVendor()	设备供应商
	getVersion()	设备版本号
	getMaximumRange()	最大取值范围

虽然表 11-1 列出的 Android 系统支持的标准传感器类型很多，但是不同厂家的 Android 系统硬件平台所支持的传感器类型不尽相同。通常情况下，在进行 Android 传感器编程时，首先要获取该 Android 系统平台所支持的传感器类型，以及传感器相关性能技术参数(如精度、量程等)。

下面编写一个程序来获取当前 Android 硬件平台上所有传感器的信息。

【例 11-1】 编写一个程序获取当前 Android 系统所支持的所有传感器类型以及传感器的相关参数。

系统主要代码如下：

```
MainActivity.java

public class MainActivity extends Activity {
    private TextView text;
    private EditText editText;
    private SensorManager mgr;
    private List<Sensor> sensors;

    protected void onCreate(Bundle savedInstanceState) {

        super.onCreate(savedInstanceState);
        setContentView(R.layout.activity_main);
        text = (TextView) findViewById(R.id.textView1);
        editText = (EditText) findViewById(R.id.editText1);
        editText.setText(getAllSensorInfoString());

    }

    // 取得传感器的类型和参数
    public String getAllSensorInfoString() {
```

```
            mgr = (SensorManager) this.getSystemService(SENSOR_SERVICE);
            sensors = mgr.getSensorList(Sensor.TYPE_ALL);
            StringBuilder message = new StringBuilder();
            text.setText("本设备内共有" + sensors.size() + "个传感器 :\n");
            int i = 0;
            for (Sensor sensor : sensors) {
                message.append("No." + (++i) + " " + sensor.getName() + "\n");
                message.append("传感器类型: " + sensor.getType() + "\n");

                message.append("制造厂商: " + sensor.getVendor() + "\n");
                message.append("系统版本: " + sensor.getVersion() + "\n");
                message.append("分辨精度: " + sensor.getResolution() + "\n");
                message.append("最大量程: " + sensor.getMaximumRange() + "\n");
                message.append("消耗功率: " + sensor.getPower() + " mA\n");
                message.append("----------------------------------\n");
            }

            return message.toString();
        }
}
```

该程序在中兴 N5 手机上的运行结果见图 11-1。

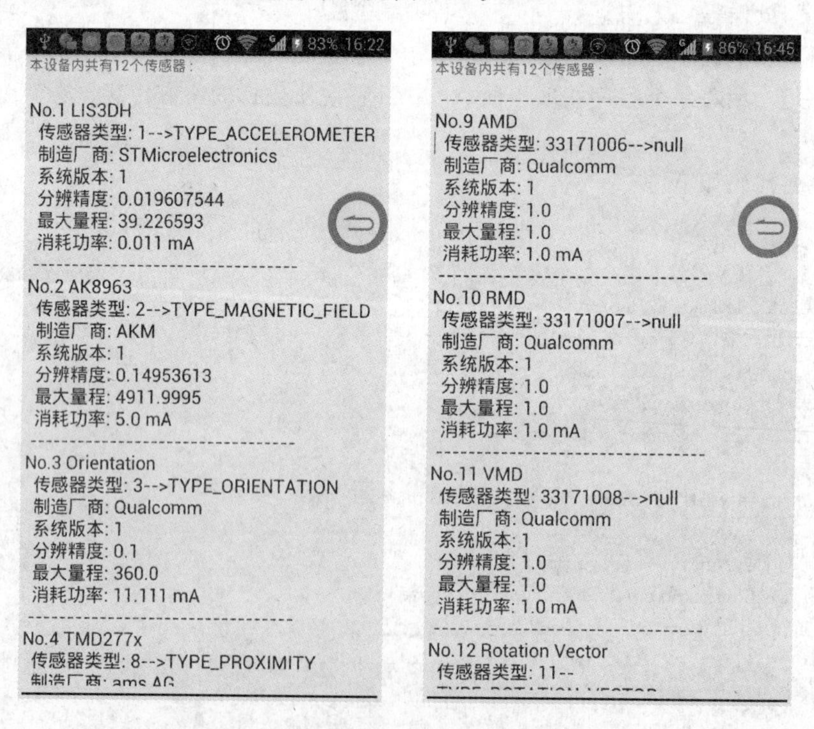

图 11-1　中兴 N5 手机上获取的传感器类型信息

从运行结果可知，该手机内置 12 种传感器，除了 Android 系统内置标准的传感器类型以外，还有厂家自行扩充的传感器类型。

11.3 Android 系统中传感器数据的采集

在获取到特定 Android 系统硬件平台中所支持的传感器信息后，我们可以从列表中选取特定传感器，采集相关传感器的数据，其步骤如下：

(1) 获取传感器管理对象 SensorManager。

调用 Context 的 getSystemService(SENSOR_SERVICE)方法获取 SensorManager。

(2) 获取指定的传感器对象 Sensor。通常有两种方法：

方法一：直接获取某种传感器的默认传感器(该类型的传感器可能不止一个，只获取并使用默认的传感器)。

```
Sensor sensor = sensorManager.getDefaultSensor(Sensor.TYPE_XXX);
```

方法二：获取某种传感器的列表(该类型的传感器可能不止一个，获取这种类型传感器的所有列表)。

```
List<Sensor> pressureSensors = sensorManager.getSensorList(Sensor.TYPE_XXX)。
```

(3) 实现 SensorEventListener 接口。

```
public interface SensorEventListener {
    public void onSensorChanged(SensorEvent event);
    public void onAccuracyChanged(Sensor sensor, int accuracy);
}
```

SensorEventListener 接口是使用传感器获取数据的关键部分，该接口包括以下两个回调函数：

onSensorChanged (SensorEvent event)方法在传感器值更改时调用。该方法只由受此应用程序监视的传感器调用。该方法的参数是 SensorEvent 对象，从该对象可以获取传感器的数值。例如，加速度传感器可以获取以下值：

```
float x = event.values[SensorManager.DATA_X];
float y = event.values[SensorManager.DATA_Y];
float z = event.values[SensorManager.DATA_Z];
```

onAccuracyChanged (Sensor sensor,int accuracy)方法在传感器的精准度发生改变时调用。该回调函数有两个参数(均为整数类型)：Sensor 表示传感器，accuracy 表示该传感器精度。

(4) 注册所要监听的传感器。应用程序要与传感器交互实现数据的采集，必须注册侦听该传感器相关的活动。为了实现该工作 SensorManager 类提供方法：

```
//注册传感器
  Boolean mRegisteredSensor = mSensorManager.registerListener(this, sensor,
  SensorManager.SENSOR_DELAY_FASTEST);
```

registerListener 方法包括 3 个参数：

第 1 个参数，接收数据传感器的 SensorEventListener 实例；

第 2 个参数，接收的传感器类型的列表(即上一步创建的 List 对象)；

第 3 个参数，接收数据的频度。

调用之后返回一个布尔值，true 表示成功，false 表示失败。

通常可以在 Activity 的 onResume()方法中，调用 SensorManager 的 registListener()为指定传感器注册监听器即可。

(5) 如果不使用该传感器了，需要将其卸载。SensorManager 类提供了方法：

```
//卸载传感器
  mSensorManager.unregisterListener(this);
```

通常可以在 Activity 的 onStop()方法中，调用 SensorManager 的 unregistListener()为指定传感器取消注册监听器即可。

11.4 加速度传感器数据的采集

接下来，我们以加速度传感器数据的采集为例，来演示 Android 应用中传感器数据的采集方法(其他类型的传感器数据采集方法与此类似，只是传感器类型和采集到的数值不同)。

Android 加速度传感器的类型是 Sensor. TYPE_ACCELEROMETER。应用程序通常是通过 android. hardware. SensorEvent 返回的加速度传感器采集值，来感知手机的运动状态或姿态数据。

该传感器采集三个参数，分别表示空间坐标系中 x、y、z 轴方向上的加速度减去重力加速度在相应轴上的分量，其单位均为 m/s2。

加速度传感器的坐标系与手机屏幕中的坐标系不同(见图 11-2)，传感器坐标系是以屏幕的左下角为原点，x 轴沿着屏幕向右，y 轴沿着屏幕向上，z 轴垂直手机屏幕向上。

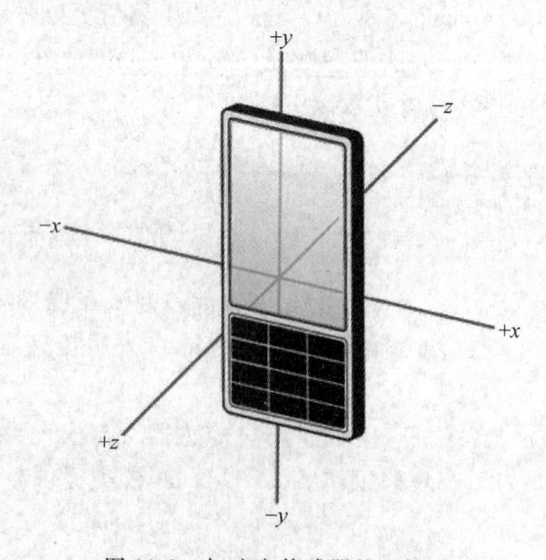

图 11-2　加速度传感器的坐标系

【例 11-2】 试编写一个程序，获取加速度传感器的动态数据并显示，并观察手机的不同位置的姿态与加速度 x、y、z 值之间的关系。

```
package com.shnu.myaccelerometer;

import android.app.Activity;
import android.hardware.Sensor;
import android.hardware.SensorEvent;
import android.hardware.SensorEventListener;
import android.hardware.SensorManager;
import android.os.Bundle;
import android.widget.TextView;

public class MainActivity extends Activity implements SensorEventListener {

    private SensorManager mSensorMgr = null;
    private Sensor mSensor = null;
    private TextView mTextView;

    public void onCreate(Bundle savedInstanceState) {
        super.onCreate(savedInstanceState);
        setContentView(R.layout.activity_main);
        mTextView = (TextView) this.findViewById(R.id.textView1);
        mSensorMgr = (SensorManager) getSystemService(SENSOR_SERVICE);
        mSensor = mSensorMgr.getDefaultSensor(Sensor.TYPE_ACCELEROMETER);
        mSensorMgr.registerListener(this, mSensor,
                SensorManager.SENSOR_DELAY_GAME);
    }


    public void onAccuracyChanged(Sensor arg0, int arg1) {

    }

    private float oldX, oldY, oldZ,total;

    public void onSensorChanged(SensorEvent arg0) {

        float x = arg0.values[0];
        float y = arg0.values[1];
        float z = arg0.values[2];
    //x,y,z 值变化太快,不容易观察,因此设了 oldX,oldY,oldZ 值便于观察
        if ((Math.abs(x - oldX)> 0.25)||(Math.abs(y - oldY)> 0.25)||(Math.abs(z - oldZ)> 0.25)) {
            oldX = x;
            oldY = y;
            oldZ = z;
            total = x + y + z;

        }
        mTextView.setText("X 轴坐标:\t" + oldX + "\t 瞬时值: " + x + "\n" +
                "Y 轴坐标:\t" + oldY + "\t 瞬时值: " + y + "\n" +
                    "Z 轴坐标:\t" + oldZ + "\t 瞬时值: " + z + "\n" +
                "Total: x + y + z = " + (total));
    }
```

通过观察手机在不同姿态下，该程序运行结果（见图 11-3 和图 11-4），基本可以得出如下结论。

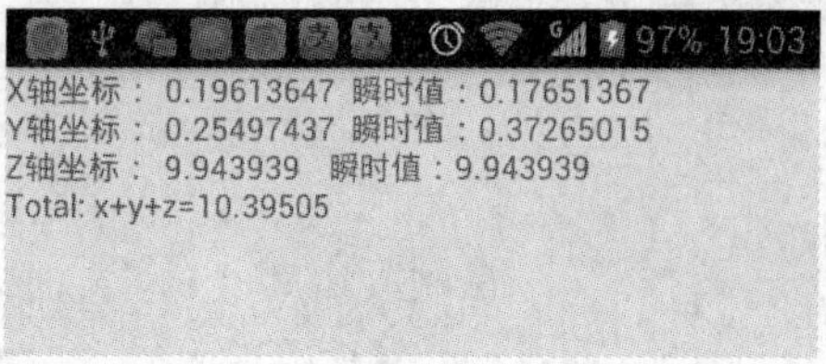

97% 19:03
X轴坐标： 0.19613647 瞬时值：0.17651367
Y轴坐标： 0.25497437 瞬时值：0.37265015
Z轴坐标： 9.943939 瞬时值：9.943939
Total: x+y+z=10.39505

图 11-3 手机水平放置时的 x,y,z 值

97% 19:05
X轴坐标： -0.05883789 瞬时值：-0.09806824
Y轴坐标： 10.081238 瞬时值：10.06163
Z轴坐标： 0.019607544 瞬时值：0.0
Total: x+y+z=10.042007

图 11-4 手机垂直放置时的 x,y,z 值

加速度传感器 Sensor. TYPE_ACCELEROMETER 的 x、y、z 轴的数值与手机的位置关系如下：

手机屏幕纵向布局、向上水平放置时(x,y,z) = (0, 0, 10)；

手机屏幕纵向布局、向下水平放置时(x,y,z) = (0, 0, −10)；

手机屏幕纵向布局、垂直放置时(x,y,z) = (0, 10, 0)；

手机屏幕纵向布局、垂直倒立放置时(x,y,z) = (0, −10, 0)；

手机屏幕横向布局、垂直放置时(x,y,z) = (10, 0, 0)；

手机屏幕横向布局、垂直倒立放置时(x,y,z) = (−10,0, 0)。

提示：可以通过在应用程序中，捕获加速度的 x,y,z 值，来判断用户是否在使用手机。

11.5 本章小结

通过本章的学习，我们已经掌握了 Android 应用程序中传感器的编程方法，学会了传感器采集相关数据的程序编写。在 Android 应用程序中，可以结合传感器，编写出用户体验更好的软件。

11.6 习题与课外阅读

11.6.1 习题

(1) 编写一个手机"掷骰子"程序，通过手机"摇一摇"，产生一个 1～6 之间的随机整数。

(2) 编写一个加速度传感器的应用，当摇动手机的时候，自动给特定的号码发送短信。

(3) 编程测试人们在行走过程中加速度传感器的状态，以及人在跌倒的时候加速度传感器的状态，据这两种状态的不同，编写一个"老人跌倒报警软件"，当老人跌倒时，Android 手机自动发报警短信，并拨通手机通讯录中亲属的电话号码。

11.6.2 课外阅读

访问下列技术网站，了解一下 Android 系统中传感器相关内容：

http://developer.android.com/reference/android/hardware/Sensor.html

第12章 网络应用

“移动”、“网络时时在线”是 Android 手机应用的一大特色，目前大部分 Android 程序的构成都有网络后台服务支撑，几乎所有的应用都涉及 Android 手机客户端与网络后台的通信。Android 的网络编程和 Java SE 的网络编程方法和 API 相差不大，它提供了 URL、TCP、UDP、蓝牙等通信 API，基于这些 API 类库，可以编写出功能丰富多彩的网络程序。

本章学习目标：

- 巩固用户线程与 UI 线程消息通信方法；
- 掌握 Android 系统网络计算模式；
- 掌握 URL、TCP、UDP 编程方法；
- 掌握 Web Service、蓝牙程序的编写方法。

12.1 网络计算模式简介

随着计算机网络和移动互联网的发展，传统上主要由 PC 完成的计算工作，已经发生了明显的变化。这个阶段的应用系统的一个明显特征是计算模式同时向两个方向分解或发展：

其一，向服务器端迁移。大量复杂的计算回归到网络后台服务器，如云计算。

其二，向移动客户端迁移。大量的应用显示与交互转向网络移动客户端，如手机、平板电脑等，移动客户端已经成为重要的开发平台。

这种计算模式既充分利用网络后台强大的计算能力，又同时拥有灵活、可移动的客户端，使用户可以随时随地使用网络后台服务。如，Android 语音识别系统应用，谷歌、百度、高德地图服务与导航系统。

Android 系统提供了 TCP、UDP 和 HTTP 的相关网络 API，这些 API 与 Java SE 的对应的 API 类完全一样，熟悉这些类库的开发人员，无须学习这部分内容，可直接上手开发网络应用。除此之外，Android 系统还集成了 Apache 的 HttpClient、WebView 控件显示网页，以及蓝牙网络通信 API。

当然，虽然网络编程相关业务逻辑代码 Android 平台和 PC 平台下完全一样，但相关数据的用户界面 UI 交互与显示是不同的，因为 Android UI 遵循单线程访问模式，对网络访问的所有应用代码必须写到 UI 线程之外的线程中，利用 Android 的消息机制实现与 UI 线程的通信，如果对此不熟悉，建议复习本书第 6 章的内容。

12.2 URL 网络程序的编写

URL 是统一资源定位符的简称，每一个 URL 对象都封装了资源的标识符和协议。如果知道网络上某个资源的 URL，就可以通过这个 URL 获取到这个资源文件，只需要按照数据流的形式读写资源即可，不必考虑协议的类型，直接按照类似文件读写数据流方式操作，就可以读写 URL 资源了。这种方式只能对 IETF(Internet Engineering Task Force)指定的 RFC 标准协议描述的资源编程。

主要代码如下：

(1) 使用 URL 获取资源。

```
URL url = new URL (urlString);
InputStream in = url.openStream();
```

(2) 获取读写 URL 数据流。

```
URLConnection conn = url.openConnection();
InputStream in = conn .getInputStream();
```

【例 12-1】 编写一个读取 URL 资源的例子。

核心代码 URLTool.java：

```
import java.io.IOException;
import java.net.MalformedURLException;
import java.net.URL;
import android.os.Bundle;
import android.os.Handler;
import android.os.Message;

public class URLTool extends Thread{
    URL url;
    Handler handler;

    public URLTool(String u, Handler handler) {
        this.handler = handler;
        try {
            url = new URL(u);     //URI.create(u).toURL();
        } catch (MalformedURLException e) {
                e.printStackTrace();
        }
    }

    public String getContent() {
        String content = "";
        try {
                content = new String(NetUtil.download(url.openConnection()));
        } catch (IOException e) {
```

```
                e.printStackTrace();
            }
            return content;
        }
        public void run(){
            String s = getContent();
            Message msg = handler.obtainMessage();
            Bundle b = new Bundle();

            b.putString("content",s);
            msg.setData(b);
            handler.sendMessage(msg);
        }
    }
```

核心代码 MainActivity.java：

```
import android.os.Bundle;
import android.os.Handler;
import android.os.Message;
import android.app.Activity;
import android.widget.EditText;

public class MainActivity extends Activity {
    EditText edText;

    @Override
    protected void onCreate(Bundle savedInstanceState) {
        super.onCreate(savedInstanceState);
        setContentView(R.layout.activity_main);
        edText = (EditText) this.findViewById(R.id.editText1);
        new URLTool("http://www.shnu.edu.cn", handler).start();

    }

    private Handler handler = new Handler() {

        public void handleMessage(Message msg) {

            Bundle b = msg.getData();
            edText.setText(b.getString("content"));
            // edText.setText(String.valueOf(msg.obj));

            super.handleMessage(msg);
        }
    };

}
```

12.3 TCP 网络编程

Android 系统支持 TCP 程序，可以编写服务器端和客户端程序，即 Android 手机既可以做服务器端，又可以做客户端。Android 网络通信相关的核心代码的编写与 Java SE 在 PC 平台下开发是一样的。

12.3.1 TCP 服务器端程序编写

TCP 服务器端程序的核心代码如下：

(1) 服务器在指定端口 PORT 上创建 ServerSocket，监听客户端的连接。

```
ServerSocket mServerSocket = new ServerSocket(PORT);
Socket socket = mServerSocket.accept();
```

(2) 连接成功后，得到返回的 socket，然后获取 socket 的输入输出流，就可以从客户端读取数据或向客户端发送数据了。

```
InputStream inStream = socket.getInputStream();
OutputStream outStream = socket.getOutputStream();
```

注意：绝大多数应用要考虑服务器与多客户端连接，服务器端程序大部分都采用多线程技术支持多客户的连接与数据交互。

12.3.2 TCP 客户端程序编写

TCP 客户端程序的核心代码如下：

(1) 与目标服务器建立连接(IP 地址、端口)。

```
Socket socket = new Socket();
SocketAddress socAddress = new InetSocketAddress(mServerIp,mPort);
socket.connect(socAddress, 5000);
```

(2) 连接成功后，从 socket 获取输入输出流，就可以从服务器端读取数据或向服务器端发送数据了。

```
InputStream inStream = socket.getInputStream();
OutputStream outStream = socket.getOutputStream();
```

12.3.3 TCP 客户端和服务器端程序编写示例

在简单复习了 TCP 服务器端和客户端的程序编写核心代码后，下面编写一个基于 Android 的 TCP 聊天程序。

【例 12-2】 编写一个程序用 TCP 实现客户端和服务端聊天。

客户端部分核心代码：

```
public class MainActivity extends Activity {
```

```
    public final static String ENCODING = "GB2312";                                  //编码方式
    private final int PORT = 2222;                                                   //连接的端口
    public static boolean STOP = true;                                               //连接是否停止
    public static final int CONNECT_SUCCESS = 0;                                     //服务器连接成功
    public static final int CONNECT_FAIL = 1;                                        // 服务器连接失败
    public static final int MESSAGE_REFRESH = 2;                                     //刷新消息界面

private EditText serverIpEditText;
    private EditText mSendEditText;
    private Button connectButton;
    private Button sendButton;
    private ListView chatListView;

    private ChatAdapter chatAdapter ;
    private List<String> chatStringList = new ArrayList<String>();  //所有聊天内容

    private Socket clientSocket;
    private ConnectThread connectThread;
    private ReceiveMessageThread receiveMessageThread;
    private OutputStream outStream;

    private Handler mHandler = new Handler(){

        @SuppressLint("HandlerLeak")
            public void handleMessage(Message msg)
        {
            switch (msg.what) {
                    case CONNECT_SUCCESS: {
                    displayToast("连接成功!");
                    serverIpEditText.setEnabled(false);
                    connectButton.setEnabled(false);
                    sendButton.setEnabled(true);
                    //开启接受线程
                    STOP = false;
                    receiveMessageThread = new ReceiveMessageThread(mHandler, clientSocket,
chatStringList);
                    receiveMessageThread.start();
                    break;
                }
                case CONNECT_FAIL: {
                        displayToast("连接失败");
                    break;
                }
                case MESSAGE_REFRESH: {
                    chatAdapter.notifyDataSetChanged();
                    break;
                }

                }
```

```
        }
    };

    @Override
    protected void onCreate(Bundle savedInstanceState) {
            super.onCreate(savedInstanceState);
            setContentView(R.layout.activity_main);
            openWifi();
            //初始化控件
            initView();
            //设置监听
            setOnListener();
        }
        private void initView() {
            // TODO Auto-generated method stub
          serverIpEditText = (EditText)this.findViewById(R.id.server_ip_text);
          mSendEditText = (EditText)this.findViewById(R.id.edittext);
          chatListView = (ListView)this.findViewById(R.id.lv_chat);
          connectButton = (Button)this.findViewById(R.id.connectbutton);
          sendButton = (Button)this.findViewById(R.id.sendbutton);
          sendButton.setEnabled(false);
          chatAdapter = new ChatAdapter(this,chatStringList);
          chatListView.setAdapter(chatAdapter);

        }

        private void setOnListener() {
            // TODO Auto-generated method stub
          //连接按钮监听
          connectButton.setOnClickListener(new View.OnClickListener()
          {
            @Override
            public void onClick(View v)
            {
                String serverIp = serverIpEditText.getText().toString().trim();
                if(!TextUtils.isEmpty(serverIp)){
                     clientSocket = new Socket();
                     connectThread = new ConnectThread(serverIp, PORT, mHandler, clientSocket);
                     connectThread.start();
                }
            }
        });

        //发送数据按钮监听
        sendButton.setOnClickListener(new View.OnClickListener()
        {

            @Override
            public void onClick(View v)
            {
```

```
            // TODO Auto - generated method stub
            byte[ ] msgBuffer = null;
            String text = "Client:" + mSendEditText.getText().toString();
            try {
                msgBuffer = text.getBytes(ENCODING);
                outStream = clientSocket.getOutputStream();
                outStream.write(msgBuffer);
            } catch (IOException e) {
                // TODO Auto - generated catch block
                Log.e("MainActivity", " ============= " + e);
            }
            chatStringList.add(text);
            chatAdapter.notifyDataSetChanged();
            //清空内容
            mSendEditText.setText("");
            displayToast("发送成功!");
        }
    });
}

@Override
public void onDestroy()
{
    super.onDestroy();
    STOP = true;
    if(clientSocket != null){
        try {
            clientSocket.close();
        } catch (IOException e) {
            e.printStackTrace();
        }
    }
    if(receiveMessageThread != null)
    {
        receiveMessageThread.interrupt();
    }
    if(connectThread != null)
    {
        connectThread.CloseClientSocket();
        connectThread.interrupt();
    }
        if(outStream != null){
        try {
            outStream.close();
        } catch (IOException e) {
            e.printStackTrace();
        }
    }
}
```

```
    //显示 Toast 函数
    private void displayToast(String s)
    {
        Toast.makeText(this, s, Toast.LENGTH_SHORT).show();
    }
        //打开 WIFI
    private void openWifi() {
        WifiManager mWifiManager = ((WifiManager) this.getSystemService("wifi"));
        if (!mWifiManager.isWifiEnabled()){
            mWifiManager.setWifiEnabled(true);
        }
    }
}
```

服务端部分代码：

```
public class MainActivity extends Activity {
    private final String ENCODING = "GB2312";                          //编码方式
    private final int PORT = 2222;                                     //连接的端口
    private boolean STOP = true;                                       //连接是否停止
    private final int CONNECT_SUCCESS = 0;                             //服务器连接成功
    private final int CONNECT_FAIL = 1;                                // 服务器连接失败
    private final int MESSAGE_REFRESH = 2;                             //刷新消息界面

    private TextView server_ip;
    private TextView ipTextView;
    private EditText mEditText;
    private Button sendButton;
    private ListView chatListView;

    public List<String> chatStringList = new ArrayList<String>();     // 所有聊天内容
    private ChatAdapter chatAdapter;
    private OutputStream outStream;
    private Socket clientSocket;
    private ServerSocket mServerSocket;

    private AcceptThread mAcceptThread;
    private ReceiveThread mReceiveThread;

    private Handler mHandler = new Handler(){
           public void handleMessage(Message msg)
      {
           switch (msg.what) {
                  case CONNECT_SUCCESS: {
                  displayToast("连接成功!");
                         ipTextView.setText("客户端的 IP 地址是: " + (msg.obj).toString());
                         sendButton.setEnabled(true);
                  break;
                  }
                  case CONNECT_FAIL: {
```

```
                        displayToast("连接失败");
                        break;
                }
                case MESSAGE_REFRESH: {
                        chatAdapter.notifyDataSetChanged();
                        break;
                }
                }
        }
};

@Override
protected void onCreate(Bundle savedInstanceState) {
        super.onCreate(savedInstanceState);
        setContentView(R.layout.activity_main);
        openWifi();
        initView();
        setOnListener();
        mAcceptThread = new AcceptThread();
        // 开启监听线程
        mAcceptThread.start();
}

private void setOnListener() {
// 发送数据按钮监听
sendButton.setOnClickListener(new View.OnClickListener() {
        @Override
        public void onClick(View v) {
        byte[] msgBuffer = null;
        String text = "Server:" + mEditText.getText().toString();
                try {
                        msgBuffer = text.getBytes(ENCODING);
                        outStream = clientSocket.getOutputStream();
                        outStream.write(msgBuffer);
                        } catch (IOException e) {
                        e.printStackTrace();
                }
                chatStringList.add(text);
                chatAdapter.notifyDataSetChanged();
                mEditText.setText("");
                displayToast("发送成功!");
        }
});
}

private void initView() {
    server_ip = (TextView) this.findViewById(R.id.server_ip);
    ipTextView = (TextView) this.findViewById(R.id.iptextview);
    mEditText = (EditText) this.findViewById(R.id.sedittext);
```

```
        sendButton = (Button) this.findViewById(R.id.sendbutton);
        sendButton.setEnabled(false);
        chatListView = (ListView) this.findViewById(R.id.lv_chat);
        chatAdapter = new ChatAdapter(this, chatStringList);
        chatListView.setAdapter(chatAdapter);
        server_ip.setText("我的 IP 地址是: " + getLocalHostIp());
    }
    @Override
    public void onDestroy() {
        super.onDestroy();
        STOP = true;
        if (mReceiveThread != null) {
                mReceiveThread.interrupt();
        }
        if (mAcceptThread != null) {
                mAcceptThread.interrupt();
        }
        if(mServerSocket != null){
                try {
                        mServerSocket.close();
                } catch (IOException e) {
                        e.printStackTrace();
                }
        }
    }
    //显示 Toast 函数
    private void displayToast(String s)
    {
        Toast.makeText(this, s, Toast.LENGTH_SHORT).show();
    }

        //打开 WIFI
    private void openWifi()
    {
        WifiManager mWifiManager = ((WifiManager) this.getSystemService("wifi"));
        if (!mWifiManager.isWifiEnabled()){
        mWifiManager.setWifiEnabled(true);
        }
    }

    // 获得本机 ip
    private String getLocalHostIp() {
        String ipaddress = "";
        try {
                Enumeration<NetworkInterface> en = NetworkInterface
                                .getNetworkInterfaces();
                while (en.hasMoreElements()) {
                        NetworkInterface nif = en.nextElement();
                        Enumeration<InetAddress> inet = nif.getInetAddresses();
```

```
                while (inet.hasMoreElements()) {
                        InetAddress ip = inet.nextElement();
                        if (!ip.isLoopbackAddress()
                                        && InetAddressUtils.isIPv4Address(ip
                                                .getHostAddress())) {
                            return ipaddress = ip.getHostAddress();
                        }
                }
            }
        } catch (SocketException e) {
            Log.e("erjiang","========= SocketException ====== " + e);
            e.printStackTrace();
        }
        return ipaddress;
    }

    // 建立连接的线程
    private class AcceptThread extends Thread {
        @Override
        public void run() {
            try {
                mServerSocket = new ServerSocket(PORT);
                clientSocket = mServerSocket.accept();
            } catch (IOException e) {
                // TODO Auto-generated catch block
                e.printStackTrace();
            }
            if (clientSocket != null) {
                Message msg = new Message();
                msg.what = CONNECT_SUCCESS;
                // 获取客户端 IP
                msg.obj = clientSocket.getInetAddress().getHostAddress();
                mHandler.sendMessage(msg);
                STOP = false;
                mReceiveThread = new ReceiveThread(clientSocket);
                mReceiveThread.start();
            }else{
                Message msg = new Message();
                msg.what = CONNECT_FAIL;
                mHandler.sendMessage(msg);
            }
        }

    }

    // 接收消息的线程
    private class ReceiveThread extends Thread {
        private InputStream mInputStream = null;
        private byte[] buf;
```

```
            private String str = null;
            ReceiveThread(Socket s) {
                try {
                mInputStream = s.getInputStream();
                } catch (IOException e) {
                    e.printStackTrace();
                }
            }
            @Override
            public void run() {
                while (!STOP) {
                    buf = new byte[512];
                    try {
                        mInputStream.read(buf);
                        str = new String(buf, ENCODING).trim();
                    } catch (IOException e) {
                        // TODO Auto-generated catch block
                        e.printStackTrace();
                    }
                    if(!"".equals(str)){
                        chatStringList.add(str);
                        Message msg = new Message();
                        msg.what = MESSAGE_REFRESH;
                        mHandler.sendMessage(msg);
                    }
                }
            }
        }
    }
```

12.4 UDP 网络编程

Android 系统下的 UDP 编程比较简单。通常 UDP 程序没有客户端和服务器端概念之分，只有发送端和接收端的区别。通信双方无须建立连接就可以发送数据，但是所发送的数据并不保证一定能送达接收端。UDP 核心的编程内容只有数据报文“发送”和“接收”两种模式。

12.4.1 UDP 数据报文的发送

向指定的目标(IP 地址和端口)发送数据。主要步骤是：

(1) 构建一个 DatagramSocket：

```
DatagramSocket udpSocket = new DatagramSocket(); //使用本地任意可用 UDP 端口
```

(2) 将要发送的数据组装成数据报文(需要知道数据报文的目标 IP 地址和端口 PORT)：

```
byte[ ] msgBuffer = text.getBytes(ENCODING);
DatagramPacket outPacket = new DatagramPacket(msgBuffer, 0,IP, PORT);
```

(3) 使用上述 DatagramSocket 发送数据包：

```
outPacket.setData(msgBuffer);
udpSocket.send(outPacket);
```

12.4.2 UDP 数据报文的接收

UDP 数据报文接收端，所要做的工作主要步骤是：

(1) 构建一个 DatagramSocket，并指明接收数据报文的端口号：

```
DatagramSocket udpSocket = new DatagramSocket(PORT);
```

(2) 构造一个空 DatagramPacket 数据报文"容器"：

```
//定义接收 UDP 网络数据的字节数组，数组长度要大于或等于对方发来的数据长度
byte[ ] inBuff = new byte[DATA_LEN];
//指定字节数组创建准备接收数据的 DatagramPacket 对象
DatagramPacket inPacket = new DatagramPacket(inBuff, inBuff.length);
```

(3) 使用上述 DatagramSocket 接收数据包：

```
udpSocket.receive(inPacket);
```

由于 UDP 通信双方通信时，须事先建立连接，通常在 UDP 数据报文接收端启动一个线程，在程序循环体(如"死循环")内来接收数据报文。

12.4.3 UDP 数据报文的发送和接收示例

在简要地复习了 UDP 数据报文的发送和接收程序编写核心代码后，下面编写一个基于 Android 的 UDP 聊天程序。

【例 12-3】 编写一个基于 Android 的 UDP 数据报文发送和接收聊天程序。

模拟的客户端代码：

```
public class MainActivity extends Activity {

    public final static String ENCODING = "GB2312";                //编码方式
    private final int PORT = 2222;                                  //连接的端口
    public static boolean STOP = true;                              //连接是否停止
//  public static final int CONNECT_SUCCESS = 0;                    //服务器连接成功
    public static final int CONNECT_FAIL = 1;                       // 数据接收异常
    public static final int MESSAGE_REFRESH = 2;                    //刷新消息界面

    private EditText serverIpEditText;
    private EditText mSendEditText;
    private Button connectButton;
    private Button sendButton;
```

```
    private ListView chatListView;

    private ChatAdapter chatAdapter ;
    private List < String > chatStringList = new ArrayList < String >();          //所有聊天内容

//  定义一个用于发送的 DatagramPacket 对象
    private DatagramPacket outPacket = null;
    private DatagramSocket clientSocket;
    private ReceiveMessageThread receiveMessageThread;

    private Handler mHandler = new Handler(){

        @SuppressLint("HandlerLeak")
        public void handleMessage(Message msg)
        {
        switch (msg.what) {
//              case CONNECT_SUCCESS: {
//              displayToast("连接成功!");
//              serverIpEditText.setEnabled(false);
//              connectButton.setEnabled(false);
//              sendButton.setEnabled(true);
//              break;
//          }
            case CONNECT_FAIL: {
                 displayToast("接收数据异常!");
                break;
            }
            case MESSAGE_REFRESH: {
                chatAdapter.notifyDataSetChanged();
                break;
            }
            }
        }
    };

    @Override
    protected void onCreate(Bundle savedInstanceState) {
        super.onCreate(savedInstanceState);
        setContentView(R.layout.activity_main);
        openWifi();
        //初始化控件
        initView();
        //设置监听
        setOnListener();
    }
    private void initView() {
        // TODO Auto - generated method stub
        serverIpEditText = (EditText)this.findViewById(R.id.server_ip_text);
        mSendEditText = (EditText)this.findViewById(R.id.edittext);
```

```
        chatListView = (ListView)this.findViewById(R.id.lv_chat);
        connectButton = (Button)this.findViewById(R.id.connectbutton);
        sendButton = (Button)this.findViewById(R.id.sendbutton);
        sendButton.setEnabled(false);
        chatAdapter = new ChatAdapter(this,chatStringList);
        chatListView.setAdapter(chatAdapter);
    }
    private void setOnListener() {
        // TODO Auto - generated method stub
        //IP 地址设置按钮监听
        connectButton.setOnClickListener(new View.OnClickListener()
        {
            @Override
            public void onClick(View v)
            {
                String serverIp = serverIpEditText.getText().toString().trim();
                if(!TextUtils.isEmpty(serverIp)){
                    try {
                            clientSocket = new DatagramSocket();
                        } catch (SocketException e) {
                            // TODO Auto - generated catch block
                            e.printStackTrace();
                        }
                    if(clientSocket != null){
                    displayToast("IP 地址设置成功!");
                    serverIpEditText.setEnabled(false);
                    connectButton.setEnabled(false);
                    sendButton.setEnabled(true);
                    //不用连接,直接监听这个数据报
                        receiveMessageThread = new ReceiveMessageThread(mHandler,
clientSocket,chatStringList);
                    STOP = false;
                    receiveMessageThread.start();
                }else{
                    displayToast("IP 地址设置失败!");
                }
            }
        }
    });

        //发送数据按钮监听
        sendButton.setOnClickListener(new View.OnClickListener()
        {
            @Override
            public void onClick(View v)
            {
                byte[] msgBuffer = null;
                String text = "Client:" + mSendEditText.getText().toString();
                try {
```

```
                //字符编码转换
                msgBuffer = text.getBytes(ENCODING);
            } catch (UnsupportedEncodingException e) {
                // TODO Auto - generated catch block
                e.printStackTrace();
            }
        // 初始化发送用的 DatagramSocket,它包含一个长度为 0 的字节数组
            try {
                    outPacket = new DatagramPacket(new byte[0], 0
    , InetAddress.getByName(serverIpEditText.getText().toString().trim()), PORT);
                } catch (UnknownHostException e) {
                    // TODO Auto - generated catch block
                    e.printStackTrace();
                }
            outPacket.setData(msgBuffer);
                // 发送数据报
            try {
                    clientSocket.send(outPacket);
                } catch (IOException e) {
                    // TODO Auto - generated catch block
                    e.printStackTrace();
                }
            chatStringList.add(text);
            chatAdapter.notifyDataSetChanged();
            mSendEditText.setText("");
            displayToast("发送成功!");
        }
    });
}

@Override
public void onDestroy()
{
    super.onDestroy();
    STOP = true;
    if(receiveMessageThread != null)
    {
    receiveMessageThread.interrupt();
    }
    if(clientSocket != null){
     clientSocket.close();
    }
}

//显示 Toast 函数
private void displayToast(String s)
{
    Toast.makeText(this, s, Toast.LENGTH_SHORT).show();
}
```

```
    //打开 WiFi
    private void openWifi() {
        WifiManager mWifiManager = ((WifiManager) this.getSystemService("wifi"));
        if (!mWifiManager.isWifiEnabled()){
            mWifiManager.setWifiEnabled(true);
        }
    }
}
```

模拟的服务端代码：

```
public class MainActivity extends Activity {

    public final static String ENCODING = "GB2312";                       // 编码方式
    private final int PORT = 2222;                                        // 连接的端口
    public static boolean STOP = true;                                    // 连接是否停止
    // public static final int CONNECT_SUCCESS = 0;                       //服务器连接成功
    public static final int CONNECT_FAIL = 1;                             // 数据接收异常
    public static final int MESSAGE_REFRESH = 2;                          // 刷新消息界面

    private TextView server_ip;
    private EditText mEditText;
    private Button sendButton;
    private ListView chatListView;

    private ChatAdapter chatAdapter;
    public List<String> chatStringList = new ArrayList<String>();  // 所有聊天内容

    private static final int DATA_LEN = 4096;
    // 定义接收网络数据的字节数组
    byte[] inBuff = new byte[DATA_LEN];
    // 以指定字节数组创建准备接收数据的 DatagramPacket 对象
    private DatagramPacket inPacket = new DatagramPacket(inBuff, inBuff.length);
    // 定义一个用于发送的 DatagramPacket 对象

    private DatagramSocket mServerSocket;
    private ReceiveMessageThread mReceiveMessageThread;
    // 定义一个用于发送的 DatagramPacket 对象
    private DatagramPacket outPacket;

    private Handler mHandler = new Handler() {

        @SuppressLint("HandlerLeak")
        public void handleMessage(Message msg) {
            switch (msg.what) {
            // case CONNECT_SUCCESS: {
            // displayToast("连接成功!");
            // serverIpEditText.setEnabled(false);
            // connectButton.setEnabled(false);
```

```
                //sendButton.setEnabled(true);
                //break;
                //}
                case CONNECT_FAIL: {
                        displayToast("接收数据异常!");
                        break;
                }
                case MESSAGE_REFRESH: {
                        sendButton.setEnabled(true);
                        chatAdapter.notifyDataSetChanged();
                        break;
                }
                }
            }
        };

        @Override
        protected void onCreate(Bundle savedInstanceState) {
            super.onCreate(savedInstanceState);
            setContentView(R.layout.activity_main);
            openWifi();
            //初始化控件
            initView();
            //设置监听
            setOnListener();
            try {
                    mServerSocket = new DatagramSocket(PORT);
            } catch (SocketException e) {
                    e.printStackTrace();
            }
            if (mServerSocket != null) {
                    mReceiveMessageThread = new ReceiveMessageThread(mServerSocket);
                    mReceiveMessageThread.start();
                    STOP = false;
            }
        }

        private void setOnListener() {
            //发送数据按钮监听
            sendButton.setOnClickListener(new View.OnClickListener() {
                    @Override
                    public void onClick(View v) {
                            //TODO Auto-generated method stub
                            byte[] msgBuffer = null;
                            String text = "Server:" + mEditText.getText().toString();
                            try {
                                    msgBuffer = text.getBytes(ENCODING);
                            } catch (UnsupportedEncodingException e) {
                                    //TODO Auto-generated catch block
```

```
                    e.printStackTrace();
                }
                try {
                    outPacket = new DatagramPacket(msgBuffer, msgBuffer.length,
                            inPacket.getSocketAddress());
                } catch (SocketException e) {
                    // TODO Auto-generated catch block
                    e.printStackTrace();
                }
                try {
                    mServerSocket.send(outPacket);
                } catch (IOException e) {
                    // TODO Auto-generated catch block
                    e.printStackTrace();
                }
                // 清空内容
                chatStringList.add(text);
                chatAdapter.notifyDataSetChanged();
                mEditText.setText("");
                displayToast("发送成功!");
            }
        });

    }
    private void initView() {
        server_ip = (TextView) this.findViewById(R.id.server_ip);
        mEditText = (EditText) this.findViewById(R.id.sedittext);
        sendButton = (Button) this.findViewById(R.id.sendbutton);
        sendButton.setEnabled(false);
        chatListView = (ListView) this.findViewById(R.id.lv_chat);
        chatAdapter = new ChatAdapter(this, chatStringList);
        chatListView.setAdapter(chatAdapter);
        server_ip.setText("我的 IP 地址是: " + getLocalHostIp());
    }

    private void openWifi() {
        WifiManager mWifiManager = ((WifiManager) this.getSystemService("wifi"));
        if (!mWifiManager.isWifiEnabled()) {
            mWifiManager.setWifiEnabled(true);
        }
    }

    // 获得本机 IP
    private String getLocalHostIp() {
        String ipaddress = "";
        try {
            Enumeration<NetworkInterface> en = NetworkInterface
                    .getNetworkInterfaces();
            while (en.hasMoreElements()) {
```

```
                    NetworkInterface nif = en.nextElement();
                    Enumeration<InetAddress> inet = nif.getInetAddresses();
                    while (inet.hasMoreElements()) {
                        InetAddress ip = inet.nextElement();
                        if (!ip.isLoopbackAddress()
                                && InetAddressUtils.isIPv4Address(ip
                                        .getHostAddress())) {
                            return ipaddress = ip.getHostAddress();
                        }
                    }
                }
        } catch (SocketException e) {
                Log.e("erjiang", "=========SocketException======" + e);
                e.printStackTrace();
        }
        return ipaddress;
    }
    //显示 Toast 函数
    private void displayToast(String s) {
        Toast.makeText(this, s, Toast.LENGTH_SHORT).show();
    }

    @Override
    public void onDestroy() {
        super.onDestroy();
        STOP = true;
        if (mReceiveMessageThread != null) {
                mReceiveMessageThread.interrupt();
        }
        if (mServerSocket != null) {
                mServerSocket.close();
        }
    }

    //接收消息的线程
    private class ReceiveMessageThread extends Thread {
        private DatagramSocket datagramSocket = null;
        private String str = null;

        ReceiveMessageThread(DatagramSocket s) {
                datagramSocket = s;
        }

        @Override
        public void run() {
                while (!STOP) {
                        try {
                                datagramSocket.receive(inPacket);
                                str = new String(inPacket.getData(), 0,
```

```
                                        inPacket.getLength(), ENCODING);
                        } catch (IOException e) {
                                Message msg = new Message();
                                msg.what = MainActivity.CONNECT_FAIL;
                                mHandler.sendMessage(msg);
                                e.printStackTrace();
                        }
                        if (!"".equals(str)) {
                                chatStringList.add(str);
                                Message msg = new Message();
                                msg.what = MESSAGE_REFRESH;
                                mHandler.sendMessage(msg);
                        }
                }
            }
        }

    }
```

12.5 HttpClient 编程

Android 的 SDK 提供了 Apache 的 HttpClient 框架以便我们使用各种 Http 服务。可以把 HttpClient 想象成一个浏览器，通过它的 Api 可以很方便地发出 GET 或 POST 请求（当然它的功能远不止这些）。

```
//创建一个默认的 HttpClient
HttpClient httpclient = new DefaultHttpClient();
//创建一个 GET 请求
HttpGet request = new HttpGet("www.google.com");
//发送 GET 请求,并将响应内容转换成字符串
HttpResponse httpResponse = httpclient.execute(request);
//取得 HttpEntiy
HttpEntity httpEntity = httpResponse.getEntity();
//通过 EntityUtils 并指定编码方式获取返回的数据
result.append(EntityUtils.toString(httpEntity, "utf-8"));
```

12.6 WebView 编程

WebKit 是一个开源浏览器网页引擎。而 android.webkit 库聚合了 Webkit 内核的浏览器功能。WebView 就是它的一个控件，可以使网页轻松地内嵌到 App 里。WebView 可以加载网页、解析 Html 语言。WebView 可以与 Javascript 互相调用。Webview 有两个方法：对于 setWebChromeClient，主要处理关于脚本的执行或 progress 等操作，而 setWebViewClient 主要处理关于页面跳转、页面请求等操作。

【例 12-4】 用 WebView 实现一个简单浏览器的功能。

```
public class MainActivity extends Activity {
    private WebView webView;
    private EditText url_edit;
    private Button go_url_btn;
    @Override
    protected void onCreate(Bundle savedInstanceState) {
        super.onCreate(savedInstanceState);
        setContentView(R.layout.activity_main);
        webView = (WebView) findViewById(R.id.message_webview);
        url_edit = (EditText)findViewById(R.id.url_edit);
        go_url_btn = (Button)findViewById(R.id.go_url_btn);


        WebSettings webSettings = webView.getSettings();
        webSettings.setJavaScriptEnabled(true);
        //解决输入框无法响应问题
        webView.requestFocusFromTouch();
        // 设置是否可缩放
        webSettings.setSupportZoom(true);
         go_url_btn.setOnClickListener(new View.OnClickListener() {
                        @Override
                        public void onClick(View v) {
                                String url = url_edit.getText().toString();
                               if(URLUtil.isNetworkUrl(url)){
                                        webView.loadUrl(url);
                                        //隐藏键盘
                                        InputMethodManager imm = (InputMethodManager)
getSystemService(Context.INPUT_METHOD_SERVICE);
                                        boolean isOpen = imm.isActive();
                                        if(isOpen){
    imm.toggleSoftInput(InputMethodManager.SHOW_IMPLICIT,
    InputMethodManager.HIDE_NOT_ALWAYS);
                                        }
                               }else{
                                        Toast.makeText(MainActivity.this, "输入网址错
误,请重新输入!", 1000).show();
                    url_edit.setText("");
                               }
                        }
                });
            //设置超链接功能
            webView.setWebViewClient(new MyWebViewClient ());
    }

     @Override
    protected void onResume() {
        super.onResume();
    }
```

```java
    //支持回退功能
    @Override
    public boolean onKeyDown(int keyCode, KeyEvent event) {
        if ((keyCode == KeyEvent.KEYCODE_BACK) && webView.canGoBack()) {
            webView.goBack();
            return true;
        }
        return super.onKeyDown(keyCode, event);
    }
      private class MyWebViewClient extends WebViewClient {
            @Override
            public boolean shouldOverrideUrlLoading(WebView view, String url) {
                view.loadUrl(url);
                return true;
            }
        }
}
```

12.7 Web Service 编程

12.7.1 Web Service 简介

根据 W3C 的定义，Web Service 是一个平台独立的、低耦合的、基于可编程的 Web 的应用程序，用于支持网络间不同机器互操作的软件系统，它是一种自包含、自描述和模块化的应用程序，使用开放的 XML 标准来描述、发布、发现、协调和配置，它可以在网络中被描述、发布和调用，可以将它看作是基于网络的、分布式的模块化组件。

Web Service 由 Web Service 提供者、Web Service 请求者、Web Service 中介者(通常是 UDDI——Universal Description Discovery and Integration 注册中心)三个角色构成，来实现 Web Service 的发布、发现、绑定使用。三个角色使用 SOAP 协议通信(见图 12-1)。

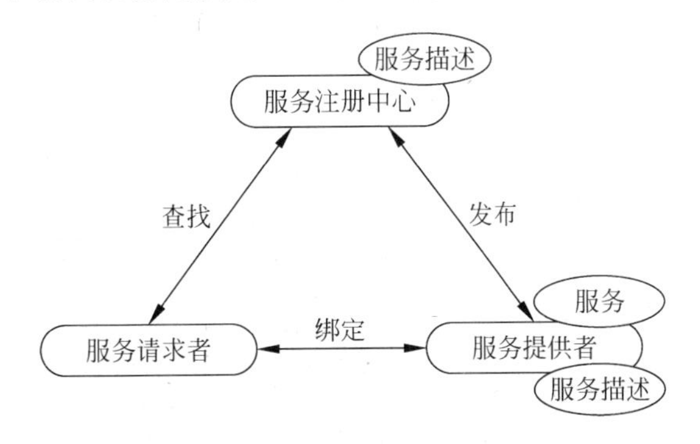

图 12-1　Web Service 三个角色之间的关系

- Web Service 提供者为其他软件提供自己已有的功能，Web Service 提供者将服务描述在 UDDI 注册。

- Web Service 请求者就是 Web 服务功能的使用者或者消费者，Web Service 请求者通过 UDDI 检索到 Web Service，并使用 Web Service。
- Web 服务中介者把一个 Web 服务请求者与合适的 Web 服务提供者联系在一起，它充当管理者的角色，一般是 UDDI。

这 3 个角色是根据逻辑关系划分的，在实际应用中，角色之间很可能有交叉：一个 Web 服务既可以是 Web 服务提供者，也可以是 Web 服务请求者，或者二者兼而有之。

Web Services 的优势在于提供了不同应用程序平台之间的互操作，使得基于组件的开发和 Web 相结合的效果达到最佳。它是基于 HTTP 协议的，调用请求和回应消息都可以穿过防火墙，不需要更改防火墙的设置，这样就避免了使用特殊端口进行通信时无法穿越防火墙的问题。

12.7.2 SOAP 协议

SOAP(Simple Object Access Protocol，简单对象访问协议)是交换数据的一种协议规范，是轻量级的、简单的、基于 XML 的用于在分布式环境中交换格式化和固化信息的简单协议。SOAP 正是 Web Service 通信中所依赖的一种协议。目前经常使用的 SOAP 协议有两个版本：SOAP 1.1 和 SOAP 1.2。

下面给出一些 SOAP 消息实例。

SOAP 请求：

```
<soapenv:Envelope
    xmlns:soapenv = "http://schemas.xmlsoap.org/soap/envelope/"
    xmlns:xsd = "http://www.w3.org/2001/XMLSchema"
    xmlns:xsi = "http://www.w3.org/2001/XMLSchema-instance">
  <soapenv:Body>
    <req:echo xmlns:req = "http://localhost:8080/wxyc/login.do">
      <req:category>classifieds</req:category>
    </req:echo>
  </soapenv:Body>
</soapenv:Envelope>
```

SOAP 请求的回应：

```
<soapenv:Envelope
    xmlns:soapenv = "http://schemas.xmlsoap.org/soap/envelope/"
    xmlns:wsa = "http://schemas.xmlsoap.org/ws/2004/08/addressing">
  <soapenv:Header>
    <wsa:ReplyTo>
      <wsa:Address> http://schemas.xmlsoap.org/ws/2004/08/addressing/role/anonymous </
wsa:Address>
    </wsa:ReplyTo>
    <wsa:From>
      <wsa:Address>http://localhost:8080/axis2/services/MyService</wsa:Address>
    </wsa:From>
    <wsa:MessageID>ECE5B3F187F29D28BC11433905662036</wsa:MessageID>
  </soapenv:Header>
```

```
    <soapenv:Body>
      <req:echo xmlns:req="http://localhost:8080/axis2/services/MyService/">
        <req:category>classifieds</req:category>
      </req:echo>
    </soapenv:Body>
</soapenv:Envelope>
```

12.7.3 WSDL 服务描述

WSDL(Web Services Description Language,Web 服务描述语言)是一种用来描述 Web 服务的 XML 语言,它描述了 Web 服务的功能、接口、参数、返回值等,便于用户绑定和调用服务。它以一种和具体语言无关的方式定义了给定 Web 服务调用和应答的相关操作和消息。

每个服务的提供都会有一个 WSDL 描述。比如,查询手机号码归属地,这个服务的 WSDL 地址是:http://webservice.webxml.com.cn/WebServices/MobileCodeWS.asmx?wsdl。访问得到的部分 WSDL 内容如下:

```
<s:schema elementFormDefault="qualified" targetNamespace="http://WebXml.com.cn/">
<s:element name="getMobileCodeInfo"><s:complexType><s:sequence>
<s:element minOccurs="0" maxOccurs="1" name="mobileCode" type="s:string"/>
<s:element minOccurs="0" maxOccurs="1" name="userID" type="s:string"/></s:sequence>
</s:complexType>
</s:element><s:element name="getMobileCodeInfoResponse">
<s:complexType>
<s:sequence><s:element minOccurs="0" maxOccurs="1" name="getMobileCodeInfoResult" type
="s:string"/>
</s:sequence></s:complexType></s:element>
<s:element name="getDatabaseInfo"><s:complexType/>
</s:element><s:element name="getDatabaseInfoResponse">
<s:complexType><s:sequence><s:element minOccurs="0" maxOccurs="1" name="
getDatabaseInfoResult" type="tns:ArrayOfString"/>
</s:sequence></s:complexType>
```

- 该 WSDL 描述 Web Service 的命名空间(NameSpace)是 http://WebXml.com.cn/;
- Web Service 提供的服务方法名称为 getMobileCodeInfo(mobileCode,userId);提供查询手机号码归属地查询,调用 getMobileCodeInfo(mobileCode,userId)方法后,返回一个名为 getMobileCodeInfoResult 的结果字符串。

12.8 Web Service 服务调用程序

(1) 首先需要第三方类库:

```
ksoap2-android-assembly-2.6.0-jar-with-dependencies.jar
```

(2) 在 AndroidManifest.xml 中添加访问网络的权限:

```
<uses-permission android:name="android.permission.INTERNET"></uses-permission>
```

(3) 组装 4 个字符串。

- 命名空间：为 WSDL 上的 http://WebXml.com.cn/。
- 调用的方法名称：为 WSDL 上的 getMobileCodeInfo。
- Endpoint：通常是将 WSDL 地址末尾的"? wsdl"去除后剩余的部分。
- SOAP Action：通常为命名空间+调用的方法名称。

(4) 设置传入参数。

在 WSDL 中能够看到调用方法需要传入的参数个数及参数名称，在设置参数时最好指明每一个传入参数的名称，如本例中的 mobileCode、userId。

(5) 读取返回结果。

返回值的类型要查看 Web 服务描述 wsdl 文档，返回值可以是键值对或数组。

如手机归属地查询 Web 服务：

```
http://webservice.webxml.com.cn/WebServices/MobileCodeWS.asmx
```

该服务提供了国内手机号码归属地查询 Web 服务，可以获取最新的国内手机号码段归属地数据，该 Web 服务提供的函数如下：

- getDatabaseInfo()

获得国内手机号码归属地数据库信息。

输入参数：无；返回数据：一维字符串数组(省份城市记录数量)。

- getMobileCodeInfo(String phoneNumber,String userID)

获得国内手机号码归属地省份、地区和手机卡类型信息。

输入参数：mobileCode =字符串(手机号码，最少前 7 位数字)，userID =字符串(商业用户 ID)免费用户为空字符串；返回数据：字符串(手机号码：省份城市手机卡类型)。

【例 12-5】 编写一个调用 Web Service 查找手机号码归属地。Web Service 资源为 http://webservice.webxml.com.cn/WebServices/MobileCodeWS.asmx。

```
WebService.java
public class WebService {

    private String ns = "http://WebXml.com.cn/";
    private String method = "getMobileCodeInfo";
    private String endPoint = "http://webservice.webxml.com.cn/WebServices/MobileCodeWS.asmx";
    private String soapAction;
    private SoapSerializationEnvelope envelope = new SoapSerializationEnvelope(
                SoapEnvelope.VER11);

    public WebService(String ns, String method, String endPoint) {
        this.ns = ns;
        this.method = method;
        this.endPoint = endPoint;
        soapAction = ns + method;
    }
    public WebService(){
        soapAction = ns + method;
        }
```

```
    public Message call(String phoneNumber) {
        phoneNumber = phoneNumber.trim();
        if (phoneNumber.length() > 11) {
                phoneNumber = phoneNumber.substring(0, 11);
        }
        // 指定 WebService 的命名空间和调用的方法名
        SoapObject rpc = new SoapObject(ns, method);
        // 设置需调用 WebService 接口需要传入的两个参数 mobileCode、userId,userID 可以不设
        rpc.addProperty("mobileCode", phoneNumber);
        //rpc.addProperty("userID", "");
        // 生成调用 WebService 方法的 SOAP 请求信息,并指定 SOAP 的版本
    envelope.bodyOut = rpc;
        // 设置是否调用的是 dotNet 开发的 WebService
        envelope.dotNet = true;
        // 等价于 envelope.bodyOut = rpc;
        envelope.setOutputSoapObject(rpc);
        HttpTransportSE transport = new HttpTransportSE(endPoint);
        try {
                // 调用 WebService
                transport.call(soapAction, envelope);
        } catch (Exception e) {
                e.printStackTrace();
        }
        // 获取返回的数据
        SoapObject object = (SoapObject) envelope.bodyIn;
        String result = null;
        // 获取返回的结果
        if(object != null){
                result = object.getProperty(0).toString();
        }else{
                result = "查询失败!";
        }
        Message message = Message.obtain();
        message.obj = result;
        return message;
    }
}
```

12.9 蓝牙通信与编程

12.9.1 蓝牙协议介绍

蓝牙(Bluetooth)这个词的来源是10世纪丹麦和挪威国王蓝牙哈拉尔(丹麦语：Harald Blåtand Gormsen),借国王的绰号“Blåtand”当名称,直接翻译成中文为“蓝牙”(blå＝蓝,tand＝牙)。蓝牙的标志是ᚼ(Hagall)和ᛒ(Bjarkan)的组合,也就是Harald Blåtand的首字母HB的合写。

蓝牙的标准是 IEEE 802.15.1，蓝牙协议工作在无须许可的 ISM（Industrial Scientific Medical）频段的 2.45GHz。最高速度可达 723.1kbps。蓝牙协议将该频段划分成 79 频道，（带宽为 1MHz）每秒的频道转换可达 1600 次。目前最新的蓝牙 4.1 版本，其优势主要体现在 3 个方面：电池续航时间、节能和设备种类，有效传输距离也有所提升，为 60m。

蓝牙协议提供点对点和点对多点的无线连接。在任意一个有效通信范围内，所有设备的地位都是平等的。蓝牙系统提供点对点连接方式（即蓝牙中仅有两点）或一点多址连接方式。在一点多址连接方式中，信道是分在几个蓝牙单元中的。分在同一信道中的两个或两个以上的单元形成一个微网（Piconet）。首先提出通信要求的设备称为主设备（Master），被动进行通信的设备称为从设备（Slave）。

- 1 台主设备最多可同时与 7 台从设备进行通信，并可以和多达 256 个从设备保持同步但不通信。
- 1 台从设备与另 1 台从设备通信的唯一途径是通过主设备转发。

利用“蓝牙”技术，能够有效地简化移动通信终端设备之间的通信，也能够成功地简化设备与 Internet 之间的通信，实现移动电话、PDA、无线耳机、笔记本电脑、相关外设等众多设备之间的无线信息交换。

12.9.2 蓝牙设备通信流程

步骤 1. 获取本机的蓝牙设备状态（如蓝牙设备是否存在，是否打开）。使用 registerReceiver 注册 BroadcastReceiver 来获取蓝牙状态、设备相关信息等。

步骤 2. 蓝牙可以扫描其他 Bluetooth 设备。通常使用 BlueAdatper 的搜索。

步骤 3. 在 BroadcastReceiver 的 onReceive()中取得搜索所得的蓝牙设备信息，如名称、MAC、RSSI。

步骤 4. 通过设备的 MAC 地址建立一个 BluetoothDevice 对象。

步骤 5. 由 BluetoothDevice 派生出 BluetoothSocket，准备 Socket 读写设备。

步骤 6. 通过 BluetoothSocket 的 createRfcommSocketToServiceRecord()方法选择连接的协议/服务。

步骤 7. Connect 之后（如果还没配对则系统自动提示），使用 BluetoothSocket 的 getInputStream()和 getOutputStream()读写蓝牙设备，见图 12-2。

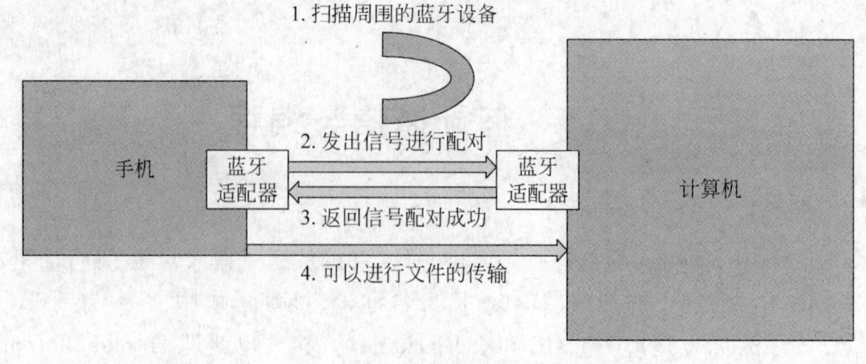

图 12-2　蓝牙通信示意图

12.9.3 蓝牙通信程序的编写

1. 蓝牙(客户端)

(1) 客户端主动连接。

```
BluetoothAdapter mBluetoothAdapter = BluetoothAdapter.getDefaultAdapter();

//根据 Bluetooth MAC address 获得 device

BluetoothDevice device = mBluetoothAdapter.getRemoteDevice(address);

//根据 UUID 创建并返回一个 BluetoothSocket,蓝牙设备管理获取远程设备

BluetoothSocket mmSocket = device.createRfcommSocketToServiceRecord(MY_UUID);

mmSocket.connect();
```

(2) 连接成功后,获取输入输出流,就可以收发信息了。

```
InputStream inStream = mmSocket.getInputStream();

OutputStream outStream = mmSocket.getOutputStream();
```

2. 蓝牙(服务器)

(1) 监听指定的端口上的客户端连接。

```
BluetoothAdapter mBluetoothAdapter = BluetoothAdapter.getDefaultAdapter();
BluetoothServerSocket mmServerSocket = mBluetoothAdapter.listenUsingRfcommWithServiceRecord
(NAME, MY_UUID);
BluetoothSocket socket = mmServerSocket.accept();
```

(2) 连接成功后,获取输入输出流,就可以收发信息了。

```
InputStream inStream = socket.getInputStream();
OutputStream outStream = socket.getOutputStream();
```

【例 12-6】 编写一个用蓝牙通信协议实现两个手机的聊天程序。

```
public class MainActivity extends Activity {
    public static final int MESSAGE_STATE_CHANGE = 1;          //监听蓝牙状态消息改变
    public static final int MESSAGE_READ = 2;                  //读消息
    public static final int MESSAGE_WRITE = 3;                 //发送消息
    public static final int MESSAGE_DEVICE_NAME = 4;           //收到远程设备的名字
    public static final int MESSAGE_TOAST = 5;                 //toast 提示
    public static final String DEVICE_NAME = "device_name";    //设备名字
    public static final String TOAST = "toast";                //bundle 的键
    private static final int REQUEST_CONNECT_DEVICE = 1;       //请求主动连接设备
    private static final int REQUEST_ENABLE_BT = 2;            //使设备可见
    private ListView mConversationView;                        //聊天记录的 listview
    private EditText mOutEditText;
```

```
    private Button mSendButton;
    private String mConnectedDeviceName;        //远程设备的名字
    private ArrayAdapter<String> mConversationArrayAdapter;
    private BluetoothAdapter mBluetoothAdapter;
    private BluetoothChatService mChatService ;

    @Override
    public void onCreate(Bundle savedInstanceState) {
        super.onCreate(savedInstanceState);
        setContentView(R.layout.activity_main);
        //获得本地蓝牙适配器
        mBluetoothAdapter = BluetoothAdapter.getDefaultAdapter();
        if (mBluetoothAdapter == null) {
            Toast.makeText(this, "该设备不支持蓝牙!", Toast.LENGTH_LONG).show();
            finish();
            return;
        }
    }

    @Override
    public void onStart() {
        super.onStart();
        // 判断蓝牙是否打开
        if (!mBluetoothAdapter.isEnabled()) {
            //跳转到蓝牙请求
            Intent enableIntent = new Intent(BluetoothAdapter.ACTION_REQUEST_ENABLE);
            startActivityForResult(enableIntent, REQUEST_ENABLE_BT);
        } else {
            if (mChatService == null) setupChat();
        }
    }

    @Override
    public synchronized void onResume() {
        super.onResume();
        if (mChatService != null) {
            if (mChatService.getState() == BluetoothChatService.STATE_NONE) {
              // 开启蓝牙聊天服务 监听外来设备来连接自己
              mChatService.start();
            }
        }
    }

    private void setupChat() {

        // 初始化聊天记录的适配器
        mConversationArrayAdapter = new ArrayAdapter<String>(this, R.layout.message);
        mConversationView = (ListView) findViewById(R.id.in);
        mConversationView.setAdapter(mConversationArrayAdapter);
```

```
        // 输入框
        mOutEditText = (EditText) findViewById(R.id.edit_text_out);
        // 发送按钮
        mSendButton = (Button) findViewById(R.id.button_send);
        mSendButton.setOnClickListener(new OnClickListener() {
            public void onClick(View v) {
                String message = mOutEditText.getText().toString();
                sendMessage(message);
            }
        });

        // 初始化蓝牙服务的线程
        mChatService = new BluetoothChatService(this, mHandler);
    }

    @Override
    public void onDestroy() {
        super.onDestroy();
        if (mChatService != null)
            mChatService.stop();
    }

    //设置本手机的蓝牙在 180 秒内可以被搜索到
    private void ensureDiscoverable() {
        if (mBluetoothAdapter.getScanMode() !=
            BluetoothAdapter.SCAN_MODE_CONNECTABLE_DISCOVERABLE) {
            Intent discoverableIntent = new Intent(BluetoothAdapter.ACTION_REQUEST_DISCOVERABLE);
            discoverableIntent.putExtra(BluetoothAdapter.EXTRA_DISCOVERABLE_DURATION, 180);
            startActivity(discoverableIntent);
        }
    }
    //发送字符串
    private void sendMessage(String message) {
        if (mChatService.getState() != BluetoothChatService.STATE_CONNECTED) {
            Toast.makeText(this,"请连接配对的设备!", Toast.LENGTH_SHORT).show();
            return;
        }
        if (message.length() > 0) {
            byte[] send = message.getBytes();
            mChatService.write(send);
            mOutEditText.setText("");
        }
    }


    private final Handler mHandler = new Handler() {
        @Override
```

```
        public void handleMessage(Message msg) {
            switch (msg.what) {
            case MESSAGE_STATE_CHANGE:
                switch (msg.arg1) {
                case BluetoothChatService.STATE_CONNECTED:
                    mConversationArrayAdapter.clear();
                    break;
                }
                break;
            case MESSAGE_WRITE:
                byte[] writeBuf = (byte[]) msg.obj;
                String writeMessage = new String(writeBuf);
                //刷新当前界面
                mConversationArrayAdapter.add("Me: " + writeMessage);
                break;
            case MESSAGE_READ:
                byte[] readBuf = (byte[]) msg.obj;
                String readMessage = new String(readBuf, 0, msg.arg1);
                mConversationArrayAdapter.add(mConnectedDeviceName + ": " + readMessage);
                break;
            case MESSAGE_DEVICE_NAME:
                mConnectedDeviceName = msg.getData().getString(DEVICE_NAME);
                Toast.makeText(getApplicationContext(), "已经连接到设备"
                                + mConnectedDeviceName, Toast.LENGTH_SHORT).show();
                break;
            case MESSAGE_TOAST:
                Toast.makeText(getApplicationContext(), msg.getData().getString(TOAST),
                               Toast.LENGTH_SHORT).show();
                break;
            }
        }
    };

    @Override
    public void onActivityResult(int requestCode, int resultCode, Intent data) {
        switch (requestCode) {
        case REQUEST_CONNECT_DEVICE:
            if (resultCode == Activity.RESULT_OK) {
                // 获取选择将要连接设备的蓝牙硬件地址
                String address = data.getExtras()
                                .getString(DeviceListActivity.EXTRA_DEVICE_ADDRESS);
                BluetoothDevice device = mBluetoothAdapter.getRemoteDevice(address);
                //开启线程,主动连接设备
                mChatService.connect(device);
            }
            break;
        case REQUEST_ENABLE_BT:
            if (resultCode == Activity.RESULT_OK) {
                    //同意请求蓝牙
                setupChat();
```

```
            } else {
                //不同意请求蓝牙
                finish();
            }
        }
    }

    @Override
    public boolean onCreateOptionsMenu(Menu menu) {
        MenuInflater inflater = getMenuInflater();
        inflater.inflate(R.menu.option_menu, menu);
        return true;
    }

    @Override
    public boolean onOptionsItemSelected(MenuItem item) {
        switch (item.getItemId()) {
        case R.id.scan:
            //连接一个设备
            Intent serverIntent = new Intent(this, DeviceListActivity.class);
            startActivityForResult(serverIntent, REQUEST_CONNECT_DEVICE);
            return true;
        case R.id.discoverable:
            //使设备可见
            ensureDiscoverable();
            return true;
        }
        return false;
    }
}
```

12.10 本章小结

通过本章的学习，已经掌握了 URL、TCP、UDP、Web Service、蓝牙协议网络程序基本编写方法，以及网络资源读写显示与 UI 线程消息通信方法。

12.11 习题与课外阅读

12.11.1 习题

(1) 编写一个基于 Android TCP 协议的文件传输程序。

(2) 编写一个基于 Android UDP 的文件传输程序。

(3) 编写一个基于 Android 蓝牙的文件传输程序。

12.11.2 课外阅读

访问下列技术网站，了解“如何在 Android 中实现一个简单连接网络的应用程序”：
http://www.chengxuyuans.com/Android/58095.html

第13章 地图导航应用

Android 手机地图导航相关软件是手机应用的一大热点，本章主要介绍基于百度 SDK 导航应用系统开发基础知识。

本章学习目标：

- 掌握百度地图 SDK 开发基本方法；
- 掌握百度地图显示与路径规划；
- 掌握路径规划与 TTS 综合应用。

13.1 百度 Android 导航 SDK 简介

百度地图 Android SDK 是一套基于 Android 2.1 及以上版本设备的应用程序接口（见表 13-1）。专注于为广大开发者提供最好的导航服务，通过使用百度导航 SDK，开发者可以轻松地为应用程序实现专业、高效、精准的导航功能。使用该套 SDK 开发适用于 Android 系统移动设备的地图应用，通过调用地图 SDK 接口，可以轻松地访问百度地图服务和数据，构建功能丰富、交互性强的地图类应用程序。百度地图 Android SDK 提供的所有服务是免费的，接口使用无次数限制。用户需在申请密钥（key）后，才可使用百度地图 Android SDK（详见 http://developer.baidu.com/map/sdk-android.htm）。

表 13-1 百度 Android SDK 提供的主要功能

主要功能	功能简介
地图功能	提供地图展示和地图操作功能。 地图展示包括普通地图（2D，3D）、卫星图和实时交通图。 地图操作：可通过接口或手势控制来实现地图的单击、双击、长按、缩放、旋转、改变视角等操作
POI 检索	支持周边检索、区域检索和城市内检索。 周边检索：以某一点为中心，指定距离为半径，根据用户输入的关键词进行 POI 检索。 区域检索：在指定的矩形区域内、根据关键词进行 POI 检索。 城市内检索：在某一城市内，根据用户输入的关键字进行 POI 检索
地理编码	提供地理坐标和地址之间相互转换。 正向地理编码：实现将中文地址或地名描述转换为地球表面上相应位置的功能。 反向地理编码：将地球表面的地址坐标转换为标准地址的过程

主要功能	功 能 简 介
线路规划	支持公交信息查询、公交换乘查询、驾车线路规划和步行路径检索。 公交信息查询：可查询公交详细信息。 公交换乘查询：根据起、终点，查询策略，进行线路规划方案。 驾车线路规划：提供不同策略，规划驾车路线(支持设置途经点)。 步行路径检索：支持步行路径的规划
地图覆盖物	支持多种地图覆盖物，展示更丰富的地图。目前所支持的地图覆盖物有定位图层、地图标注(Marker)、几何图形(点、折线、弧线、多边形等)、地形图图层、POI检索结果覆盖物、线路规划结果覆盖物等
	导航地图控制：放大、缩小、2D视角、3D视角。 导航信息展示：转向标、路线信息、指南针、道路信息、GPS信号强弱、电子眼、限速播报、比例尺等
定位功能	采用GPS、WiFi、基站、IP混合定位模式，使用Android定位SDK可获取定位信息，使用地图SDK定位图层进行位置展示
离线地图	可以通过手动和SDK接口两种形式导入离线地图包，使用离线地图可节省用户流量，提供更好的地图展示效果
导航功能	目前SDK支持调启百度地图客户端导航和调启Web页面导航(H5导航)(注意：调启百度地图导航需要设备提前安装5.0及以上版本的百度地图)
LBS云	是百度地图针对LBS开发者全新推出的平台级服务，不仅适用PC应用开发，同时适用移动设备应用的开发。使用LBS云，可以实现移动开发者存储海量位置数据的服务器零成本并缓解维护压力，且支持高效检索用户数据，且实现地图展现
特色功能	特色功能包括短串分享、Place详情页展示等。 短串分享：将POI搜索结果或反地理编码结果生成短串，当其他用户点击短串即可打开手机上的百度地图客户端或者手机浏览器进行查看； Place详情页：以可视化界面展示POI搜索的详细信息

13.2 开发环境配置

百度Android导航SDK是一套基于Android 2.1及以上版本设备的应用程序接口，可以通过该接口实现专业的导航功能。主要的导航部分软件开发工作有：

(1) 路线规划开发——通过输入起点、途经点和终点，可以发起路线规划。

(2) 导航功能开发——成功发起路线规划后，可以进入导航。百度导航支持真实GPS导航、模拟导航、文字导航和HUD导航。

(3) 语音播报——SDK分为集成TTS模块与非集成TTS模块两种版本：集成了TTS模块的版本适用于第三方应用中未集成TTS模块，且在导航中需要语音播报的功能；未集成TTS模块的版本适用于第三方应用中已经集成了其他TTS模块。

开发者可在百度Android导航SDK的下载页面下载到最新版的导航SDK，下载地址为http://developer.baidu.com/map/navsdk-android-download.htm。

为了给开发者带来更优质的导航服务、满足开发者灵活使用SDK的需求，百度导航SDK提供了两个下载包——集成了TTS模块的版本和未集成TTS模块版本。使用不带语音播报的 SDK，开发者可以使用自己的 TTS 去播报导航过程中语音文本。

13.2.1 申请密钥

为了给用户提供更安全优质的服务，LBS 开放平台针对 Android 平台的 SDK 产品引入 Key 认证机制，用户在使用之前需要先申请配置 Key，并在程序相应位置填写自己的 Key。

Key 机制：每个 Key 仅且唯一对于一个应用验证有效，即对该 Key 配置环节中使用的包名匹配的应用有效，因此，多个应用(包括多个包名)需申请多个 Key，或者对一个 Key 进行多次配置。

Key 的申请地址为：http://lbsyun.baidu.com/apiconsole/key。

说明：若需要在同一个工程中同时使用导航 SDK、定位 SDK 和地图 SDK，可以共用同一个 Key。因为不同机器的签名不同，如果想运行案例中的 DEMO，请自行到官网申请自己的 Key。

13.2.2 SDK 开发环境配置

百度导航 SDK 由 3 部分组成：代码 Jar 包、资源包和 so 动态库(见图 13-1)。

- 代码 Jar 包：由 Java 源代码编译打包而成，提供在线导航、线路规划、语音播报等功能。
- 资源包：由导航所需的配置数据、基础数据以及导航功能所需的 layout、drawable、string 等资源打包而成。
- so 动态库：是由 native 代码编译而成，主要是地图、导航、路线规划、语音播报等功能的底层实现。

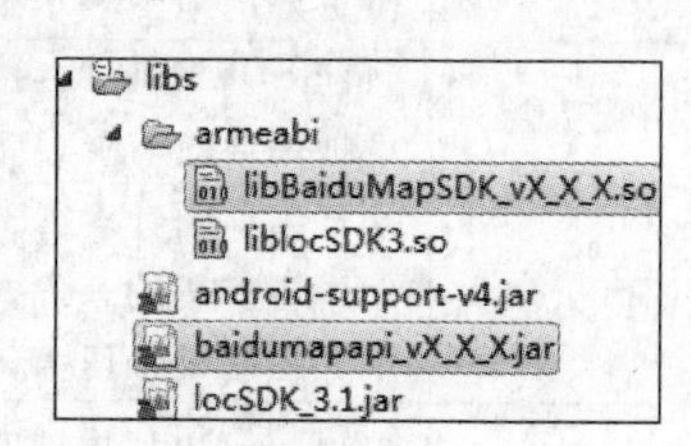

图 13-1 百度 SDK 目录

上述 3 部分 SDK 开发环境配置工作如下：

(1) 在新建的 Android 工程里的 assets 目录添加 BaiduNaviSDK_Resource_vX_X_X.png 和 channel；

(2) 在 libs 目录添加 BaiduNaviSDK_vX_X_X.jar、android_api_1.1_forsdk.jar、need_lib.jar；

(3) 在 libs/armeabi 目录添加 libapp_BaiduNaviApplib_vX_X_X.so、libejTTS.so、libCNPackage.so。

注意：对于 need_lib.jar，它是百度移动统计 SDK 的部分，如果开发者同时也使用了百度移动统计 SDK，并且添加 need_lib.jar 到工程时候发生代码冲突，则应该把 need_lib.jar 去除掉。

13.3 开发工作步骤

百度导航 SDK 提供便捷的方式让开发者快速发起导航：

(1) 创建并配置开发环境。

(2) 在 AndroidManifest.xml 中添加使用百度导航所需权限及 Android 版本支持，在 application 中添加开发密钥。

```
<application
    <meta-data
        android:name="com.baidu.lbsapi.API_KEY"
        android:value="开发者 key" />
</application>
```

导航需要的权限如下：

```
<uses-permission android:name="android.permission.GET_ACCOUNTS" />
<uses-permission android:name="android.permission.USE_CREDENTIALS" />
<uses-permission android:name="android.permission.MANAGE_ACCOUNTS" />
<uses-permission android:name="android.permission.AUTHENTICATE_ACCOUNTS" />
<uses-permission android:name="android.permission.ACCESS_NETWORK_STATE" />
<uses-permission android:name="android.permission.INTERNET" />
<uses-permission android:name="com.android.launcher.permission.READ_SETTINGS" />
<uses-permission android:name="android.permission.CHANGE_WIFI_STATE" />
<uses-permission android:name="android.permission.ACCESS_WIFI_STATE" />
<uses-permission android:name="android.permission.READ_PHONE_STATE" />
<uses-permission android:name="android.permission.WRITE_EXTERNAL_STORAGE" />
<uses-permission android:name="android.permission.BROADCAST_STICKY" />
<uses-permission android:name="android.permission.WRITE_SETTINGS" />
<uses-permission android:name="android.permission.READ_PHONE_STATE" />
```

（3）初始化导航引擎。

在应用的入口 activity 中设置代码如下：

```
private boolean mIsEngineInitSuccess = false;
private NaviEngineInitListener mNaviEngineInitListener = new NaviEngineInitListener() {
                    public void engineInitSuccess() {
                        //导航初始化是异步的,需要一小段时间,以这个标志来识别
//引擎是否初始化成功,为 true 时候才能发起导航
                        mIsEngineInitSuccess = true;
                    }
                    public void engineInitStart() {
                    }

                    public void engineInitFail() {
                    }
                };
            private String getSdcardDir() {
                    if (Environment.getExternalStorageState().equalsIgnoreCase(
                            Environment.MEDIA_MOUNTED)) {
                        return Environment.getExternalStorageDirectory().toString();
                    }
                    return null;
                }
            public void onCreate(Bundle saveInstance) {
                    super.onCreate(saveInstance);
                    //初始化导航引擎
```

```
                        BaiduNaviManager.getInstance().
                            initEngine(this, getSdcardDir(), mNaviEngineInitListener,
"我的 key",null);
                    }
```

（4）配置导航页 activity。

再新建一个 activity，在 Manifest 中加入导航页的声明。

```
<activity android:name = ".BNavigatorActivity"
    android:configChanges = "orientation|screenSize|keyboard|keyboardHidden"/>
```

初始化导航页 activity：

```
//导航监听器
        private IBNavigatorListener mBNavigatorListener = new IBNavigatorListener() {

                @Override
                public void onYawingRequestSuccess() {
                    // TODO 偏航请求成功

                }

                @Override
                public void onYawingRequestStart() {
                    // TODO 开始偏航请求

                }

                @Override
                public void onPageJump(int jumpTiming, Object arg) {
                    // TODO 页面跳转回调
                    if(IBNavigatorListener.PAGE_JUMP_WHEN_GUIDE_END == jumpTiming){
                        finish();
                    }else if(IBNavigatorListener.PAGE_JUMP_WHEN_ROUTE_PLAN_FAIL ==
jumpTiming){
                        finish();
                    }
                }

                @Override
                public void notifyGPSStatusData(int arg0) {

                }

                @Override
                public void notifyLoacteData(LocData arg0) {

                }
```

```
            @Override
            public void notifyNmeaData(String arg0) {

            }

            @Override
            public void notifySensorData(SensorData arg0) {

            }

            @Override
            public void notifyStartNav() {
                BaiduNaviManager.getInstance().dismissWaitProgressDialog();
            }

            @Override
            public void notifyViewModeChanged(int arg0) {
                // TODO Auto - generated method stub

            }

        };

    public void onCreate(Bundle savedInstanceState){
            super.onCreate(savedInstanceState);

            //创建 NmapView
            MapGLSurfaceView nMapView = BaiduNaviManager.getInstance().createNMapView
(this);

            //创建导航视图
             View navigatorView = BNavigator.getInstance().init(BNavigatorActivity.
this, getIntent().getExtras(), nMapView);

            //填充视图
            setContentView(navigatorView);

            BNavigator.getInstance().setListener(mBNavigatorListener);
            BNavigator.getInstance().startNav();

            // 初始化 TTS. 开发者也可以使用独立 TTS 模块,不用使用导航 SDK 提供的 TTS
            BNTTSPlayer.initPlayer();

            //设置 TTS 播放回调
            BNavigatorTTSPlayer.setTTSPlayerListener(new IBNTTSPlayerListener() {

                @Override
                public int playTTSText(String arg0, int arg1) {
                    //开发者可以使用其他 TTS 的 API
```

```
                    return BNTTSPlayer.playTTSText(arg0, arg1);
                }

                @Override
                public void phoneHangUp() {
                    //手机挂断
                }

                @Override
                public void phoneCalling() {
                    //通话中
                }

                @Override
                public int getTTSState() {
                    //开发者可以使用其他 TTS 的 API
                    return BNTTSPlayer.getTTSState();
                }
            });

             BNRoutePlaner.getInstance().setObserver(new RoutePlanObserver(this, new
IJumpToDownloadListener() {

                @Override
                public void onJumpToDownloadOfflineData() {

                }
            }));

        }

        @Override
            public void onResume() {
                BNavigator.getInstance().resume();
                super.onResume();
                BNMapController.getInstance().onResume();
            };

            @Override
            public void onPause() {
                BNavigator.getInstance().pause();
                super.onPause();
                BNMapController.getInstance().onPause();
            }

            @Override
            public void onConfigurationChanged(Configuration newConfig) {
                BNavigator.getInstance().onConfigurationChanged(newConfig);
                super.onConfigurationChanged(newConfig);
```

```
        }

        public void onBackPressed(){
            BNavigator.getInstance().onBackPressed();
        }

        @Override
        public void onDestroy(){
            BNavigator.destory();
            BNRoutePlaner.getInstance().setObserver(null);
            super.onDestroy();
        }
```

(5) 在导航初始化成功后，在第三个 activity RouteGuideDemo 中输入起始点，发起导航。

```
BaiduNaviManager.getInstance().launchNavigator(this,
        40.05087, 116.30142,"百度大厦",
        39.90882, 116.39750,"北京天安门",
        NE_RoutePlan_Mode.ROUTE_PLAN_MOD_MIN_TIME,              //算路方式
        true,                                                   //真实导航
        BaiduNaviManager.STRATEGY_FORCE_ONLINE_PRIORITY,        //在离线策略
        new OnStartNavigationListener() {                       //跳转监听
            @Override
            public void onJumpToNavigator(Bundle configParams) {
                Intent intent = new Intent(RouteGuideDemo.this, BNavigatorActivity.class);
                intent.putExtras(configParams);
                startActivity(intent);
            }

            @Override
            public void onJumpToDownloader() {
            }
        });
```

13.4 导航功能开发

13.4.1 简介

算路成功后会获得算路结果 RoutePlanModel，然后即可以根据算路结果发起导航，导航方式分为模拟导航和真实 GPS 导航两种。进入到模拟导航或者 GPS 导航后，单击转向标按钮即可以切换到文字导航，在文字导航界面，可以切换到 HUD 模式。

13.4.2 配置导航页 activity

新建一个 activity，在 AndroidManifest.xml 中加入导航页的声明。

```
<activity
    android:name=".BNavigatorActivity"
    android:configChanges="orientation|screenSize|keyboard|keyboardHidden"/>
```


初始化导航页 activity：

```
//导航监听器
private IBNavigatorListener mBNavigatorListener = new IBNavigatorListener() {

    @Override
    public void onYawingRequestSuccess() {
        // TODO 偏航请求成功

    }

    @Override
    public void onYawingRequestStart() {
        // TODO 开始偏航请求

    }

    @Override
    public void onPageJump(int jumpTiming, Object arg) {
        // TODO 页面跳转回调
        if(IBNavigatorListener.PAGE_JUMP_WHEN_GUIDE_END == jumpTiming){
         finish();
        }else if(IBNavigatorListener.PAGE_JUMP_WHEN_ROUTE_PLAN_FAIL == jumpTiming){
            finish();
        }
    }

    @Override
    public void notifyGPSStatusData(int arg0) {

    }

    @Override
    public void notifyLoacteData(LocData arg0) {

    }

    @Override
    public void notifyNmeaData(String arg0) {

    }

    @Override
    public void notifySensorData(SensorData arg0) {

    }
```

```
    @Override
    public void notifyStartNav() {
        BaiduNaviManager.getInstance().dismissWaitProgressDialog();
    }

    @Override
    public void notifyViewModeChanged(int arg0) {

    }

};

public void onCreate(Bundle savedInstanceState){
    super.onCreate(savedInstanceState);

    //创建 NmapView
    MapGLSurfaceView nMapView = BaiduNaviManager.getInstance().createNMapView(this);

    //创建导航视图
    View navigatorView = BNavigator.getInstance().init(BNavigatorActivity.this, getIntent
().getExtras(), nMapView);

    //填充视图
    setContentView(navigatorView);

    BNavigator.getInstance().setListener(mBNavigatorListener);
    BNavigator.getInstance().startNav();

    // 初始化 TTS. 开发者也可以使用独立 TTS 模块,不用使用导航 SDK 提供的 TTS
    BNTTSPlayer.initPlayer();

    //设置 TTS 播放回调
    BNavigatorTTSPlayer.setTTSPlayerListener(new IBNTTSPlayerListener() {

        @Override
        public int playTTSText(String arg0, int arg1) {
            //开发者可以使用其他 TTS 的 API
            return BNTTSPlayer.playTTSText(arg0, arg1);
        }

        @Override
        public void phoneHangUp() {
            //手机挂断
        }

        @Override
        public void phoneCalling() {
            //通话中
```

```
            }

            @Override
            public int getTTSState() {
                //开发者可以使用其他 TTS 的 API,
                return BNTTSPlayer.getTTSState();
            }
        });

        BNRoutePlaner.getInstance().setObserver(new RoutePlanObserver(this,
                                                    new IJumpToDownloadListener() {

            @Override
            public void onJumpToDownloadOfflineData() {

            }
        }));

    }


    @Override
    public void onResume() {
        BNavigator.getInstance().resume();
        super.onResume();
        BNMapController.getInstance().onResume();
    };

    @Override
    public void onPause() {
        BNavigator.getInstance().pause();
        super.onPause();
        BNMapController.getInstance().onPause();
    }

    @Override
    public void onConfigurationChanged(Configuration newConfig) {
        BNavigator.getInstance().onConfigurationChanged(newConfig);
        super.onConfigurationChanged(newConfig);
    }

    public void onBackPressed(){
        BNavigator.getInstance().onBackPressed();
    }

    @Override
    public void onDestroy(){
        BNavigator.destory();
        BNRoutePlaner.getInstance().setObserver(null);
        super.onDestroy();
    }
```

13.4.3 发起导航

下面的算路结果 mRoutePlanModel 是算路成功后获取的,详细内容可参考路线规划指南。

```
private void startNavi(boolean isReal) {
if (mRoutePlanModel == null) {
        Toast.makeText(this, "请先算路!", Toast.LENGTH_LONG).show();
        return;
}
// 获取路线规划结果起点
RoutePlanNode startNode = mRoutePlanModel.getStartNode();
// 获取路线规划结果终点
RoutePlanNode endNode = mRoutePlanModel.getEndNode();
if (null == startNode || null == endNode) {
        return;
}
// 获取路线规划算路模式
int calcMode = BNRoutePlaner.getInstance().getCalcMode();
Bundle bundle = new Bundle();
bundle.putInt(BNavConfig.KEY_ROUTEGUIDE_VIEW_MODE,
            BNavigator.CONFIG_VIEW_MODE_INFLATE_MAP);
bundle.putInt(BNavConfig.KEY_ROUTEGUIDE_CALCROUTE_DONE,
            BNavigator.CONFIG_CLACROUTE_DONE);
bundle.putInt(BNavConfig.KEY_ROUTEGUIDE_START_X,
            startNode.getLongitudeE6());
bundle.putInt(BNavConfig.KEY_ROUTEGUIDE_START_Y,
            startNode.getLatitudeE6());
bundle.putInt(BNavConfig.KEY_ROUTEGUIDE_END_X, endNode.getLongitudeE6());
bundle.putInt(BNavConfig.KEY_ROUTEGUIDE_END_Y, endNode.getLatitudeE6());
bundle.putString(BNavConfig.KEY_ROUTEGUIDE_START_NAME,
            mRoutePlanModel.getStartName(this, false));
bundle.putString(BNavConfig.KEY_ROUTEGUIDE_END_NAME,
            mRoutePlanModel.getEndName(this, false));
bundle.putInt(BNavConfig.KEY_ROUTEGUIDE_CALCROUTE_MODE, calcMode);
if (!isReal) {
        // 模拟导航
        bundle.putInt(BNavConfig.KEY_ROUTEGUIDE_LOCATE_MODE,
                    RGLocationMode.NE_Locate_Mode_RouteDemoGPS);
} else {
        // GPS 导航
        bundle.putInt(BNavConfig.KEY_ROUTEGUIDE_LOCATE_MODE,
                    RGLocationMode.NE_Locate_Mode_GPS);
}

Intent intent = new Intent(RoutePlanDemo.this, BNavigatorActivity.class);
intent.putExtras(bundle);
startActivity(intent);
}
```

发起模拟导航：startNavi(false)；

发起 GPS 导航：startNavi(true)；

百度地图导航的功能已经实现完毕，要想了解更多关于语音播报功能和路径规划功能，请参照官方文档。

13.5 本章小结

通过本章的学习，我们已经掌握了 Android 系统下基于百度导航系统的应用程序编写，具备了进一步编写基于百度导航系统的 LBS 应用开发能力。

13.6 习题与课外阅读

13.6.1 习题

(1) 简述百度地图导航应用开发步骤。

(2) 基于百度地图 SDK 开发一个计算两地空间距离的软件。

13.6.2 课外阅读

访问下列技术网站，了解 Baidu LBS Android 应用方面的应用：

http://developer.baidu.com/map/

参考文献

[1] Darcey, L. S. Conder. Android Wireless Application Development Volume I: Android Essentials, Addison-Wesley,2012.

[2] Ekong, D. U. A survey of android programming courses. Southeastcon, 2012 Proceedings of IEEE, 2012.

[3] Lee, W.-M. Beginning android 4 application Development, John Wiley & Sons,2012.

[4] Meier, R. Professional Android 4 application development, John Wiley & Sons,2012.

[5] Milette, G. A. Stroud. Professional Android sensor programming, John Wiley & Sons,2012.

[6] Rogers, R. , et al. Android application development: Programming with the Google SDK, O'Reilly Media, Inc,2012.

[7] 郭志宏. Android应用开发详解[M]. 北京：电子工业出版社,2010.

[8] 杨丰盛. Android应用开发揭秘[M]. 北京：机械工业出版社,2010.

[9] 余志龙,陈昱勋,郑名杰,陈小凤,郭秩均. Google Android SDK开发范例大全[M]. 北京：人民邮电出版社,2009.

[10] 靳岩,姚尚朗. Google Android开发入门与实践[M]. 北京：人民邮电出版社,2009.

[11] E2ECloud工作室. 深入浅出Google Android[M]. 北京：人民邮电出版社,2009.